THOUGHTS ON RELIGION

THOUGHTS ON RELIGION

KEVIN A. FOX

First Edition

2018
ISBN 978-1-947587-01-4

Front Cover: Galileo's middle finger.
Adapted from publicity photograph from Museo Galileo, Florence.

Book design and composition by Katie Fox

San Francisco
FOXEDITING.COM

CONTENTS

DEDICATION

I want to dedicate this little book to those closest to me—my loving companion and better half, Rosalinda (Lyn) Taylor and my three, most remarkable children—Katie, Colleen and Sam. To Lyn, because she has always supported my writing endeavors and has patiently waited—coat on, purse in hand—to go out for lunch while I finish writing down some "interesting final" thoughts that later will prove to be not really very interesting and certainly not final. She has served as First Proofreader on most of my earlier writing, but was spared that chore this time. To my children, because they are an unending source of pride— all three are especially interesting in their various ways, and all would be chosen BFFs if not already so by blood. Thanks for all you do, guys!

ACKNOWLEDGMENTS

I am again honored to thank my elder daughter, Katie Fox, for her exemplary editorial skills and for employing her enormous breadth of knowledge to redirect my thinking to more relevant/ productive discussions. I would not be able to pull this all together without her guidance. Her ability to diplomatically point out the mundane errors of grammar and expression (and even spelling) generated by my nearly eighty-year-old brain is exceptional. I'm not sure where she acquired such patience— certainly not from me!

As noted in the FOREWORD, this rambling effort summarizes some of my many thoughts on religion over several decades. It is by no means a work of scholarship—it only reviews what I perceive to be major failures of our principal religions and examines whether the merits of belief can make up for those failures. That said, I remind the reader that I alone am responsible for any errors of content and interpretation herein. I confess my expertise in theology and philosophy is limited, but hope that my largely science-based perception of religion offsets those limitations.

FOREWORD

"Religion teaches people to be extremely self-centered and conceited. It assures them that god cares for them individually, and it claims that the cosmos was created with them specifically in mind." —Christopher Hitchens

"I am against religion because it teaches us to be satisfied with not understanding the world." —Richard Dawkins

"The unexamined life is not worth living." —Socrates

In June 2017 the global opinion research group Ipsos released results of their carefully executed survey of Canadians' views on religion. Remarkably, a small but significant majority of Canadians indicated that they believed *"religions did more harm than good."* Still more remarkably, residents of Québec, traditionally the most religious province, led the nation in their disapproval of religion's effects (56 percent vs. 51 percent national average). Canadians often exhibit more progressive views than Americans, but many religious "scholars" were still surprised by the results. I expect that if Ipsos were to survey the world's other enlightened and progressive nations they would get similar results—citizens of Japan, Australia, and New Zealand, as well as Great Britain, France, Germany and other northern European nations, would likely share the Canadian perspective that religion has a negative effect on society.

Despite their modest disapproval of religion in general, nine out of ten Canadians said they were *"completely comfortable being around people who have different religious beliefs than me."* Canadians would seem to be mostly tolerant—yet 13 percent (18 percent of Québécois!) said they lose respect for someone when they find out they are religious. Of course, 13–18 percent is small compared to the opposite response by the 30 percent of Americans who lose respect for others when they find they are atheists!

In an earlier (2011) survey, Ipsos established that 32 percent of Canadians believed that religious people were better citizens. By 2017 that view has declined to only 24 percent, but

Canadian atheists still experience a social stigma. In contrast, Daniel Cox of the Public Religion Research Institute (2017) advises us that about a third of Americans think atheists should be banned from higher public office and not be allowed to teach in our public schools. American atheists have been routinely stigmatized by public figures including Presidents (e.g., Reagan, Bush I and Trump—in his inaugural address, Trump spoke of bringing all Americans together with "the same almighty Creator"). That stigma distorts participant responses in all but the most carefully constructed surveys of religious attitudes—traditional measures of belief in America have grossly underestimated the number of atheists. Many if not most atheists conceal their belief in telephone and mail surveys as well as in face-to-face interviews. The most recent carefully executed survey by Will Gervais and Maxine Najle (2017) indicates that it is likely that 26 percent of Americans do not believe in god—that is in contrast to traditional survey estimates of 3–10 percent. That number may be even higher—no surveys that I know of clearly define "God." They fail to distinguish between "a personal God" (an omnipotent anthropomorphic entity which the believer actively worships) and "an unknowable god" (a universal creative force that is beyond human comprehension). If survey subjects are thinking of the latter they may be unable to say they are atheists even though they may not believe in a personal God. Dan Cox at the Public Religion Research Institute suggests that only about half of Americans surveyed *have the anthropomorphic God in mind, while for other believers it's something far more abstract.* It is very likely that more Americans are non-believers than we previously thought.

These issues of belief continue to intrigue me—I have read and thought about religious "purpose" often in my life, and for more than a decade I have been making notes on the scientific basis of human belief and on the contributions of religion to human culture. I don't pretend to be a religious scholar, but I am capable of reading and understanding research articles in most areas of the natural and social sciences. My perspective is principally scientific/rational and not theological/speculative.

I originally wrote this little book to refine and organize my own thoughts on the merits of religion in contemporary culture, but more recently have realized I should probably share them while I'm alive in order to discuss and defend them. My intended audience is mainly my children, other close relatives, and friends. Others should feel free to read it but must recognize that my progressive, secular worldview does not conform to the traditional and often-conservative viewpoints expressed by other members of my family.

Please be forewarned—I am trained as a scientist and I'm not very tolerant of the irrationality that prevails in today's Western culture—nor do I have much patience with those who don't examine the basis of their own beliefs (Socrates' *"unexamined life"*). I fear that humankind's gift of reason is so widely neglected today that people worldwide feel empowered to believe anything that makes them feel superior and/or more secure. Our American society has become increasingly ignorant over the past four or five decades and now lags educationally and intellectually behind almost all of the world's advanced societies. Many Americans lack any real understanding of social, natural and physical science. They are unable to rationally evaluate evidence to arrive at reasoned conclusions, and increasingly, they allow scientifically illiterate media talk-show hosts and/or religious charlatans to provide them with ready-made worldviews. It is no wonder that political and religious extremism thrives in today's culture—we become ever more polarized as in our efforts to be "open-minded" (and "politically correct") we patiently tolerate increasingly bizarre and irrational beliefs. Perhaps we should be less gracious and follow Thomas Jefferson's advice about irrational belief: *"Ridicule is the only weapon which can be used against unintelligible propositions."*

I agree with the contemporary American adage that *"everyone is entitled to their own opinion"*—I just expect that it should be an informed and rational opinion. Unfortunately, a significant number of Americans hold patently false, even stupid, beliefs on a wide range of critically important global issues. Let me cite just three examples.

One: *"Evolution is just a theory and is no more valid than the Biblical view of Creation."*

If you are unable to distinguish between "truths" based on faith and those based upon reason you will probably agree with this statement. This "anti-Darwinian" perspective reveals a woeful lack of understanding of scientific methods and knowledge. It reflects the anti-scientific agenda promoted by certain Christian and Muslim extremists who fear the gradual erosion of their personal theologies and the power of their particular supreme being. For more than five centuries, scientific progress in the West has continued to diminish the creator's role despite repeated efforts by "the Church" to suppress new knowledge. (Galileo's middle finger still gestures defiantly at Urban VIII, the pope who labeled him a heretic and persecuted him for endorsing the Copernican view that the Earth revolves around the Sun.)[1]

Contrary to George W. Bush's assertion that *"on the issue of evolution, the verdict is still out in how God created the Earth,"* the matter is fully resolved among rational, informed and educated people worldwide. Modern evolutionary theory enables humankind to enjoy a clear view of our place in the universe and is the principal explanatory paradigm behind most recent advances in social, biological, medical and pharmaceutical sciences. Most attempts to discredit evolutionary theory come from fundamentalist theologians and are ultimately based upon fear and ignorance. If you don't understand evolution I urge you to either read a good book on the subject or just shut up and don't pretend that you do. If you are capable of reading objective, secular, non-fiction, please consider Jerry Coyne's *Why Evolution is True* (Penguin, 2009) or Richard Dawkins' *The Greatest Show on Earth—The Evidence for Evolution* (Free Press, 2009).

Two: *"The planet Earth is not overpopulated and can support many billions more people—there are no real food shortages, only food distribution problems."* You've heard this repeatedly—similar arguments have been traditionally made for expanding U.S. immigration—e.g. *"There is plenty of vacant land in the Western states."* If millions of Americans continue to believe this, they may be condemning your great-great-grandchildren to death from starvation! Every day, one out of every seven people goes to bed hungry, and about 30–35,000 people, mostly children, die of malnutrition and hunger-related diseases. It will continue to worsen—I remember learning as a youngster that the food resources of the oceans were inexhaustible—we've certainly done a pretty good job of depleting/downgrading those resources in the last 50 years.

This problem is a little more complex than the previous one. First you must understand that much of the food we produce to feed the planet's more than 7 billion inhabitants comes from unsustainable fossil-fuel-subsidized agriculture. The "Green Revolution" following WWII permitted us to greatly expand global food production. Norman Borlaug, the American plant pathologist who was the guiding genius behind this "agricultural miracle," won the 1970 Nobel Prize for his work. Following WWII there was a worldwide excess of nitrate-based munitions that were easily converted to inexpensive fertilizer. Borlaug's approach was very practical—he simply selected the crop strains that were most responsive to the then-cheap nitrate fertilizers. Semi-barren grasslands could be fertilized and irrigated to produce high yields of important cereal grains. Strains of rice that responded best to nitrate fertilizer were selected for, and the yield per acre doubled. Similar results were seen with wheat and other food crops. Before he died at age 95 in 2009, Borlaug openly acknowledged that his efforts to eliminate global famine were an ineffective stopgap measure that may, in fact, have created greater long-term problems for humankind.

Commercial fertilizers contain nitrogen either as nitrate, ammonia, or both in the form of ammonium nitrate. But unlike the post-war years when we could use old munitions, today we manufacture virtually all of these nitrogen components using fossil fuel sources. Fossil fuels serve as both the feedstock and the energy source for the chemical reactions in the manufacturing processes. In the simplest terms, it takes about a ton of natural gas to make a ton of ammonia fertilizer. (Students of organic chemistry may recall the "Haber process" that combines hydrogen from natural gas with nitrogen from the air to make ammonia). In his later years, Norman Borlaug agreed with the assertion that today we are doing little more than *"converting fossil fuel into food."* We all should know that fossil fuels are a finite, non-renewable and rapidly diminishing resource. We should also know that expanded use of fossil fuels in the last century has had profound effects on our atmosphere and global climate.

Global dependence on fossil-fuel-subsidized agriculture increases every year. Western atmospheric scientists express concern about China's growing number of coal-fueled generating plants but neglect to point out that much of that energy will be used to synthesize the fertilizer needed to sustain the agricultural base to feed her more than one billion people. Even if we have enormously "good luck" with genetically engineered agriculture, we will still encounter insurmountable problems feeding the world's population in the next several generations. Natural disasters, war, famine and pestilence may be the only "solutions." It should be obvious to anyone that the death tolls from these four very real "Apocalyptic horsemen" increase as population increases. Nobody dies from "overpopulation" *per se*—they die from the social and environmental consequences of it. The enormous loss of life during the 2004 Indonesian tsunami and the 2010 Haitian earthquake should be recognized as largely the result of overpopulation! Too many people are compelled to live in marginally habitable locations. Similarly, the spread of most infectious diseases increases in direct proportion to population density.

By 1970, when Borlaug won his Nobel Prize, it had become apparent to most scientists that the global population was passing a critical threshold. In order to sustain ourselves we would have to exploit and even deplete many non-renewable resources and irreversibly damage the earth's ecosystems. Amazingly, since that year we have allowed the global population to double (from ~3.6 billion in 1970 to ~7.2 billion in 2017). Global population will exceed 9 billion by 2050 and 10 billion by 2085. Unfortunately, our global food production has increased only slightly since 1990 and is not likely to increase enough to fully accommodate those increased numbers.

By the second half of this century, China's per capita agricultural production is projected to decline as her population surges past 1.5 billion. Similar crises will likely occur worldwide. Our grandchildren will witness major global food shortages and environmental deterioration by the end of the twenty-first century. And by the next century humankind can expect uncontrolled massive emigrations by environmental, nutritional and economic refugees from over-populated Third World countries. If you have been paying attention, you know that it has already started! From 2000 to 2010, Syria had one of the highest population growth rates in the world. The revolution in that country was largely precipitated by the terrible drought of 2006 to 2011 that killed 85 percent of the livestock in western Syria and forced over a million people to abandon their farms. Appeals to President Bashar al-Assad fell on deaf ears, the environmental refugees sought assistance in the already overcrowded cities, and the rest is history.

Throughout the Middle East and Africa, the enormous population momentum will overwhelm the educational and social support systems, and uneducated, unemployable, angry young people will flood the planet. A preview of those societies can be seen in rural Pakistan today, where social support systems are almost nonexistent and less than half the children attend school. Of those that do go to "school," about half attend the madrasas where they do little more than memorize Islamic scripture and learn to hate the West.

Why have we done nothing about human population growth? Religion, politics and economics—it is political suicide for an elected official to advocate population control in a religion-dominated culture that denies there is any problem. Most fundamentalist religions continue to promote large families. In the US, nearly all politicians respond by promoting population growth. We still reward the lack of family planning (i.e., large families) with overly generous dependency allowances and unlimited child tax credits. Capitalists know that an excessive labor force drives down wages and production costs.

Failure to regulate immigration to the US and Europe has encouraged citizens of Central and South America, Asia, Africa and the Middle East to continue their unchecked reproduction and to *"export their excess offspring to the West."* The world is running out of food and habitable space—much of that *"vacant land"* in North America is either unsuitable for human habitation or is a source of limited renewable resources (e.g., timber, grassland, water, etc.) critical to our survival. Extreme optimists and deniers still believe that new technologies for energy production will continue to save us, but they fail to consider all the unintended consequences of the technology required to feed ten or more billion people.[2] Population crisis deniers often argue that today's areas of explosive human population growth will pass through some modern version of the "demographic transition"—as these regions emerge from pre-industrial to industrial economic systems their birth rates will decline to levels comparable to western Europe today. Unfortunately, there is little good evidence that the Demographic Transition model is remotely applicable to the less economically developed countries of Africa and the Middle East. The model was based principally upon the changes in Europe, North America and Japan over the past several centuries—today's developing countries don't have two or more centuries to complete such a transition.

Unchecked population growth may be the single greatest threat confronting our species—perhaps even greater than global climate change.

Three: *"Global climate change is a hoax."* Sorry folks, but even George W. Bush eventually acknowledged the reality of climate change. Dick Cheney and Rush Limbaugh will probably go to their graves insisting that anthropogenic climate change is a conspiracy by *"power-hungry environmentalists"* (whomever they might be). Our current President, Donald Trump, claims it is a hoax perpetrated by the Chinese to disrupt our economy. The sad truth is that the many politicians who deny climate change do so because they personally have something to lose if it becomes widely accepted, not because they have any real understanding of the problem. This controversy disturbs me greatly because so many presumably educated people are weighing in on an issue about which they know almost nothing.

I know many academics, from economists to philosophers, who *choose* not to believe in global climate change even though they have not the faintest understanding of the science involved. I never understood that accepting scientific findings was still a matter of ideological choice or faith—but apparently in the US and the Middle East that is still the case. I am not an atmospheric scientist, but I do understand the data and have read much of the climate change literature published in reputable journals—I have no doubt that anthropogenic global climate change is occurring and if unchecked, will have devastating effects on our descendants. But I am especially alarmed by the number of uninformed people taking strong positions on issues they don't understand. When were the most ignorant Americans encouraged to vote on the validity of scientific theories? Perhaps back about the time Ronald Reagan announced that he had *"serious doubts about the theory of evolution."* Despite its frequent use as a rhetorical device, personal incredulity does not constitute evidence! Rejecting an idea just because you *"find it hard to believe"* only shows that you lack understanding (are ignorant) and/or are incapable of understanding (are stupid). Out of politeness, we will give Reagan the benefit of the doubt and assume he never mastered the science required to have an informed opinion. Of course, in that case he should have kept his mouth shut!

We agree with those who argue that Earth has experienced greater temperature fluctuations in the past. But we qualify it by saying those changes were in the remote past, long before the appearance of modern *Homo sapiens,* and we emphasize that most of those changes took place over tens of thousands, not hundreds, of years. If the global warming deniers are not silenced we may soon pass the "tipping point"—over the next several centuries our planet's warming process will become irreversible, and human life as we currently know it will cease to exist. But the planet may be a better place without the contemporary form of human life!

So, dear reader, if you don't believe in evolution, don't think that there is a global population crisis and you're sure that global climate change is a hoax, I suggest you do not attempt to read this little book. I am convinced that the cultural milieu that promotes such irrational thinking by Americans and others is largely the result of today's pervasive religious fundamentalism. In fact, I hold organized religions responsible for many if not most of our global problems. The discussion that follows examines humankind's dependence on religion and the relative merits and dangers of that dependence. If you are a very religious, devout theist you may find much of that material offensive. If that is the case you might want to stop reading at this point. To borrow a line from Daniel Dennett: *"Others are advised to close the book now and tiptoe away!"*

For those of you who plan to bravely read on, let me urge you to take time to read the Notes. If you have read my other works you know that I tend to enrich and further document the narrative in my notes—they are often more interesting than the text itself. Because many of them are quite lengthy we have left them at the end, rather than as footnotes at the bottom of the page. Just read the book with a bookmark in the Notes to make it easy for you to flip back to the appropriate page. Have fun reading, and I hope you will enjoy at least some of my thoughts.

My Non-Belief System

"The invisible and the non-existent look pretty much alike to me." —*Anonymous*

"He who possesses art and science has religion; he who does not possess them, needs religion." —*Goethe*

"Atheism is nothing more than a commitment to the most basic standard of intellectual honesty: one's convictions should be proportional to one's evidence. Pretending to be certain when one isn't—indeed, pretending to be certain about propositions for which no evidence is even conceivable—is both an intellectual and a moral failing." —*Sam Harris*

In September of 2010, the Pew Forum released the results of a survey on religious knowledge by belief group. The results of their "U.S. Religious Knowledge Survey" were widely reported in the popular press. Participants were asked questions on "Bible and Christianity," "World Religions," and "Religion in Public Life." Self-described "atheists/agnostics" exhibited significantly greater overall knowledge of religion than did the "very religious"—Roman Catholics and Evangelical Christians scored lowest. White Evangelicals scored fairly well on their knowledge of the Bible but little else. The success of atheists comes as no surprise when you acknowledge that they are, on average, more intelligent than believers. If you don't like that assertion please read on—we fully document it later.

I developed my secular worldview and philosophy quite early. I used to describe myself as a "Christian atheist," but that usually required some explanation. In recent years, the term "Christian" has been appropriated by the fundamentalist right, and to them, "Christian atheist" is an oxymoron.[3] But of course, for many of us, "Christian" simply means a follower of Christ's basic teachings. In truth, our study of the "historical Jesus" shows we know very little for certain about him and his original teachings, or for that matter, if he ever actually existed. Some scholars feel that he is nothing more than an amalgam of pre-Christian heroes and myths. Still, most agree that his reputed teachings came to emphasize love, tolerance and forgiveness, and they represented an abrupt departure from the earlier harshly retributive codes of the Sumerians (Ur-Nammu) and Babylonians (Hammurabi) as well as those of the Old Testament. Richard Dawkins has recently described that god of the Old Testament as *"a misogynistic, homophobic, racist, infanticidal, genocidal, megalomaniacal, sadomasochistic, capriciously malevolent bully."* If you are offended by Dawkins' characterization I urge you to carefully read your Old Testament—it's all there for you!

I began questioning the existence of God in early adolescence. My mother had warned us of a vengeful god who would punish us for our sins. As a child, anytime something bad happened to me I believed it was punishment from God. Mother professed that the death of my sister Christine Elizabeth (aka Christina, Christena, "Tena") was God's punishment for her (Mother) having married a non-Catholic. She had told me when I was very young, perhaps four or five, that my father would be sent to Hell because he was never properly baptized and she prevailed upon me to plead with him to seek salvation through baptism. I recall tearfully asking Dad one evening to please be baptized Catholic so he could go to Heaven with the rest of us. He said nothing—he just looked at me, and Mother standing behind, with a gravely pained expression. As I grew older I realized that a just god would not consign someone to Hell for "failing to register" as a Catholic!

My mother did whatever she could to keep her children Catholic. I recall her urging me to stay a safe distance from the church behind our house on Livermore Street because they were *"crazy people."* At first I wondered if they might kidnap and brainwash me, but soon gave it little thought. I later discovered that it was a Unitarian-Universalist Church—I guess in Mother's eyes they were a little crazy not to embrace her Catholic god. For several years she required my brother Neal and me to read Bible stories on Saturday mornings when she could hold us captive—a sort of coerced, home-schooled, Catholic catechism. Our Aunt Margaret supplied the instructional materials. I actually enjoyed reading about Jesus and the early Christians. We were a "Christian," if not a Catholic, household. Dad was a practicing Christian in his fundamental goodness and generosity. I also think he was very much a theist—perhaps stronger in his belief than Mother herself. One time in a moment of great candor, Mother admitted to me that she did not really know whether there was a god or an afterlife. I guess it was her attempt to show me the value of "faith." Years after her death I carefully read her extensive writings and realized that indeed, her faith was far from absolute.

ASKING THE FORBIDDEN QUESTION

"When I was a child, I spake as a child, I understood as a child, I thought as a child; but when I became a man, I put away childish things." —Saint Paul, I Corinthians 13:11

(The apocryphal, second-century "Acts of Paul" describes the saint as "a rotund man of small stature, with a bald head and bandy legs ... with large eyes ... eyebrows meeting and nose somewhat hooked ... and the face of an angel." That doesn't sound like an angel to me. Like many people, I enjoyed reading Paul for the poetic beauty of the King James Bible, but unlike most, I didn't take his often superior and intolerant rhetoric very seriously. In fact, I have always enjoyed perverting Paul's famous quote above to describe my own maturation from a fearful childish belief to a healthy adult atheism—I have "put away childish things." I expect that bald, bow-legged, chubby, little Jew would turn over in his grave if he knew I was taking such liberties with his prose.)

I remember the first time I was able to seriously talk at length about the existence of God with another person. That was a taboo subject in our house, and most of my Catholic friends wouldn't dare to seriously entertain such thoughts. They knew that if they ever left their Church they would go to Hell; my few Protestant friends were equally certain of the unique redemptive value of their own belief systems. I wondered just how many times you could be sent to hell for failing to find the "true Faith"—Homer Simpson, that greatest of American philosophers, may have said it best: *"Suppose we've chosen the wrong god. Every time we go to church we're just making him madder and madder."*

My first serious discussion about God's existence was with Billy Abbott, two years my senior and a classmate of my brother Neal. Billy was member of a relatively affluent and prominent local family. He and I shared an interest in science and math, and we both remained informed on contemporary scientific topics. I remember little of the conversation except that we both agreed that if many of our childhood heroes such as Poe, Twain, Darwin, Edison, Einstein, Freud, Hemingway and Asimov were atheists (and they surely were, despite efforts by revisionist historians to assert otherwise), then there might be something to it. I was 13 or 14 at the time and it was a clearly an epiphany (or perhaps more appropriately, an *anti-epiphany*)—I had actually been able to express my doubts regarding God and had not been ridiculed or threatened. I saw no need to continue to embrace that greatest of human myths, but I also knew that I could never share my opinions widely. That early "loss of faith" did not involve a corresponding loss of moral compass. I was still enamored with the life of Jesus and found his message compelling, even if I found little use for the peripheral myths, rituals and archaic laws. At that time, however, I did not fully understand that the moral values I was developing were neither uniquely "Christian" nor entirely learned.

In the pages that follow I will frequently refer to the "Abrahamic faiths"—meaning the three principal Semitic monotheistic traditions of Judaism, Christianity and Islam. These religions all trace their roots to the covenant God made with Abraham as described in Genesis. This is the pact which assured the Jews that they were "the chosen people" and granted them all the land from the Nile to the Euphrates.[4] Today most objective religious scholars agree that the Abrahamic religions were not created *de novo* by God and Abraham, but instead have antecedents in other, centuries-older cultures. For example, analysis of Egyptian writings from the 14th century BC reveal that many of the myths and several hymns believed to be original to Judaism were part of earlier Egyptian culture. Today many objective scholars believe that the central tenets of Judaism, and by extension, Christianity and Islam, originated in Egypt. In the 14th century BC, the Pharaoh Akhenaten and his "divinely" beautiful Queen, Nefertiti, founded a new "monotheistic" religion. It has been suggested that the earliest ancestors of today's Jews were living in Egypt at that time and embraced the new faith. Later, when Akhenaten and Nefertiti's new religion was abandoned by the majority of the Egyptians, those resident Jews who remained faithful left Egypt as religious dissidents. The story of Exodus tells us they were led out by Moses and founded the monotheistic Abrahamic tradition in the ancient nation of Israel. The "interesting parallels" between the ancient Egyptian gods and myths and the accounts in the Judeo-Christian Bible are far too numerous to be due to coincidence. Similarly, there are striking similarities between sixth-century BC Babylonian Creation and Great Flood myths and the comparable narratives in Genesis 1–11. The evidence suggests that many Biblical accounts are derived from the older Babylonian and Egyptian mythologies. As always, it is difficult to distinguish mythology from ancient history.

The more we study the original Judeo-Christian divine texts, the more we are convinced that they are neither "original" nor "divine." They are instead a remarkable collection of pre-existing myths, civil codes and historical accounts that have been blended and reworked into often compelling narratives. There is evidence that these ancient, "divinely inerrant," sacred texts were extensively edited by mortal men, and we continue to rediscover ancient texts that were apparently judged inappropriate and discarded by various church authorities through the ages. For example, recent findings suggest that the Old Testament was extensively edited to eliminate references to Yahweh's wife, the fertility goddess Asherah, who apparently figured prominently in Jewish religious practice prior to the eighth-century BC. References to Asherah are still found in the Book of Kings and other ancient texts. Numerous figurines, amulets and pottery inscriptions honoring her attest to her status as an embarrassing "second" deity in a supposedly monotheistic belief system. (He didn't say, "*Thou shalt have no other gods before me and my wife.*")

Of course the New Testament suffered a similar fate at the hands of religious editors. Constantine the Great (AD 272–337), the first Roman Emperor to convert to Christianity, took a great interest in church doctrine and convened the first Ecumenical Council of the Catholic Church (aka "The First Council at Nicaea"). In addition to promulgating such profound canon law as the prohibition of self-castration (see below) and disallowing clerics from cohabiting with the young single women (the so-called *virgenes subintroductae* who ostensibly served as their domestic aides), the Council did unify Christian doctrine. There is little doubt that those texts that conformed to the early doctrine were embraced and those that did not were either edited or discarded.

There are many other problems with scriptural validity—the "Synoptic Problem" for Biblical scholars refers to the "uncanny" similarities between the gospels of Mark, Matthew and Luke. These Synoptic Gospels, which are the principal source of our "knowledge" about Jesus, include many of the same stories in the same order, often with the exact same wording. Most scholars agree that Mark is the oldest and that the author(s) of Matthew and Luke used Mark as a source. In addition, some Biblical scholars hypothesize the existence of a lost gospel, "Q," to explain the considerable material (~25 percent) shared in common by Matthew and Luke but not found in Mark. April DeConick, Professor of Biblical Studies at Rice University, advises us that the "Synoptic Problem" is *"unresolvable"* because *"our manuscript tradition is at best third-century ... we have no idea what the Gospel of Mark actually said in the first century, or the Gospels of Luke or Matthew. We might act like we do. But the truth is we don't."* Dr. DeConick's work reminds us that the transmission process over the first three centuries was not just simple copying, but relied heavily on oral recitation and fallible human memory.

Tragically, many people believe that these "sacred texts" are the crucial roots of our morality and it is our duty to obey all their instructions. A more reasonable interpretation is that some of these texts contain the civil codes of ancient peoples, and often lack relevance to contemporary cultures. As Westerners we often assume that moral definitions are an integral part of all religions. That is simply not so—Greek and Roman religions contained no moral injunctions. People got their moral standards from the prevailing philosophy and/or from community practices. Few Eastern religions have anything equivalent to the rigorous moral commands of the Abrahamic faiths. Confucianism, which continues to guide hundreds of millions of Asians, is a *non-religion* that emphasizes ethical/

moral life without requiring belief in the supernatural. In a similar fashion, millions of contemporary secular humanists live moral lives based on ethics and reason rather than faith—and millions of Buddhists live meaningful and moral lives without belief in a supreme being. The inescapable truth is that human morality is neither derived from nor dependent upon supernatural religious belief. In fact, there is considerable evidence that supernatural belief often perverts morality.

ATHEIST OR AGNOSTIC—DOES IT REALLY MATTER?

"I contend we are both atheists; I just believe in one fewer god than you do. When you understand why you dismiss all the other possible gods, you will understand why I dismiss yours."
—*Stephen F. Roberts*

"Militant Muslims become suicide bombers; militant Christians blow up abortion clinics; militant atheists write books." —*Anonymous*

"I am an atheist, out and out. It took me a long time to say it. I've been an atheist for years and years, but somehow I felt it was intellectually unrespectable to say one was an atheist, because it assumed knowledge that one didn't have. Somehow, it was better to say one was a humanist or an agnostic."
—*Isaac Asimov*

The word agnostic simply means "to not know." It is simple and accurate—an agnostic contends that the existence of God is unknowable that he personally doesn't know whether God exists or not. (It is actually possible to be an agnostic theist—one who believes in an unknowable God!) In contrast, the word atheistic means "without a belief in God." Here is a subtle but important distinction—an atheist does not by definition *deny* the existence of God—she/he simply does not have that belief. Atheism itself is not a belief; it is the absence of belief. To lack a belief in God only means that the entity has no relevance in one's world view. Admittedly, some atheists go beyond the simple lack of belief and actively argue against the existence of God, but that is not a defining quality for being an atheist.

I generally call myself an atheist because it suggests to others that I am firm in my convictions. If I say I'm an agnostic they will assume I have reservations and that I might be interested in the nonsense they will marshal in support of their particular god. As a scientist I know that I can't disprove a non-entity, so I can never really be an "absolute" atheist. On the other hand, Bertrand Russell's famous "Celestial Teapot" analogy beautifully refutes the notion that the burden of proof is on the skeptic (me) to disprove the believers' scientifically untestable claims. In his 1952 essay, "Is There a God?" Russell playfully argued:

> *If I were to suggest that between the Earth and Mars there is a china teapot revolving about the sun in an elliptical orbit, nobody would be able to disprove my assertion provided I were careful to add that the teapot is too small to be revealed even by our most powerful telescopes. But if I were to go on to say that, since my assertion cannot be disproved, it is an intolerable presumption on the part of human reason to doubt it, I should rightly be thought to be talking nonsense. If, however, the existence of such a teapot were affirmed in ancient books, taught as the sacred truth every Sunday, and instilled into the minds of children at school, hesitation to believe in its existence would become a mark of eccentricity and entitle the doubter to the attentions of the psychiatrist in an enlightened age or of the Inquisitor in an earlier time.*

I don't believe in either Russell's teapot or the god of Abraham—neither proposition is supported by any objective evidence, and my worldview is clearer unencumbered by such irrelevancies. My training as a scientist predisposes me to atheism. As British geneticist J.B.S. Haldane explained more than 50 years ago: *"My practice as a scientist is atheistic. That is to say, when I set up an experiment I assume that no god, angel or devil is going to interfere with its course; and this assumption has been justified by such success as I have achieved in my professional career. I should therefore be intellectually dishonest if I were not also atheistic in the affairs of the world."*

Atheism/agnosticism is widespread among scientists—in a widely reported recent survey by sociologist Elaine Ecklund (*Science vs. Religion: What Scientists Really Think*, Oxford University Press, 2010), 64 percent of scientists clearly identified themselves as nonbelievers. The fact that about a quarter of those atheists/agnostics also described themselves as "spiritual" confused many readers who naïvely assumed that such individuals must really be "believers." That is, of course, nonsense—one doesn't have to believe in god to be spiritual—ask any Buddhist! Like many of my fellow scientists, I consider myself a spiritual atheist! I am compelled to point out that

Ecklund's research was and continues to be funded by the John Templeton Foundation, which has strong ties to fundamentalist Protestantism. Furthermore, a significant number of reviewers were critical of Ecklund's inclusion of many fellow social scientists in her survey—not everyone agrees that sociologists, economists, political scientists etc. are representative "scientists." Despite the many criticisms, Ecklund's work is still important. My admittedly preliminary examination of her data suggests that 70 percent might be a more accurate estimate of atheism/ agnosticism among natural scientists. A 2009 Pew Research Center survey gave comparable results—only 33 percent of American Association for the Advancement of Science (AAAS) members professed belief in a personal god. AAAS includes a diverse group of scientists many with little background in the natural and physical sciences.

More significantly, comparable surveys of the *most prominent* scientists reveal that an overwhelming majority are atheists/ agnostics. Only about seven percent of the members of the National Academy of Sciences claim to believe in a personal god. The corresponding value among Fellows in the Royal Society, the UK's equivalent organization, is only about five percent. Careful studies of belief and religiosity among science Nobel laureates reveal similar results.

Why is this so? You may suppose that it is indoctrination in the methods of science that causes scientists to reject matters of faith. While that may be a contributing factor, the overwhelming evidence suggests that religiosity and belief in God are negatively correlated with intelligence. That is, people of higher intelligence, whether scientifically inclined or not, tend to be less religious. The fact that Nobel laureates in *non-science* disciplines (Literature, Economics, Peace) also tend to be atheists bears this out. Numerous surveys show that atheists constitute a high proportion of the membership of the various high-IQ societies—73 percent of Mensa members (eligibility requires an IQ in the top two percent) and 98 percent of Intertel members (IQ in top one percent) do not profess any religious beliefs. Both organizations include a majority of nonscientists. (In the interest of full disclosure, I confess I have for many years claimed membership in both groups.)

The distinguished American psychologist Albert Ellis once said, only somewhat facetiously: *"There is a very obvious negative correlation between religiosity and intelligence. The only thing we are not sure about is whether only stupid people become religious, or whether religion makes them stupid."*

A recent study (Lindeman and Svedholm-Häkkinen, "Does Poor Understanding of Physical World Predict Religious and Paranormal Beliefs?"; *Applied Cognitive Psychology*, June 2016) further confirms this perspective—poor reasoning skills and ignorance of basic principles of physics and biology appear to be major factors contributing to belief in God. In the authors' words: *"Strong religious beliefs are correlated with poor intuitive physics skills, poor mechanical ability, poor mental rotation, low school grades in mathematics and physics, poor common knowledge about physical and biological phenomena, (poor) intuitive and analytical thinking styles. ...*

By now I expect that I've offended some more readers—I'm sorry if you are one. You may want to stop reading here—it will only get worse! If you are not offended and want to read more on this topic you might start with Satoshi Kanazawa's 2010 now-classic paper entitled *"Why Liberals and Atheists are More Intelligent"* (*Soc. Psych. Quart.* Vol. 73, No. 1, 33–57).

Many of the famous people whom I've admired are atheists. I often didn't know they were atheists when I first became impressed with their work, and in some cases I had been led to believe they were actually theists. For centuries many believers have sought to establish that various great men were really theists despite evidence to the contrary. Every few years another revisionist "historian" will come forward to claim that Einstein or some other great person was really a theist. Apparently, these are the same people who regularly claim to have knowledge of deathbed conversions by Thomas Paine, Mark Twain and Charles Darwin that we know never happened. Einstein, like many other scientists, often used the word *"God"* as a metaphor for the unknown. Many scientists use terms such as *"the God particle,"* *"the eye of God"* or *"God's mind"* to characterize the furthest frontiers of human understanding. Theists have repeatedly seized on Einstein's metaphorical use of the word "god" as evidence that he was a believer. The fact that he sometimes described himself as *"a deeply religious nonbeliever"* apparently confused many theists hoping to claim him as one of their own. In 1954 an exasperated Einstein was compelled to write a letter defining his position. He emphatically stated: *"It was, of course, a lie what you read about my religious convictions, a lie which is being systematically repeated. I do not believe in a personal God and I have never denied this but have expressed it clearly. If something is in me which can be called religious then it is the unbounded admiration for the structure of the world so far as our science can reveal it."* That same year Einstein also wrote: *"The word God is for me nothing more than the expression and product of human weaknesses, the Bible a collection of honourable but still primitive legends which are nevertheless pretty childish. No interpretation no matter how subtle can (for me) change this."* Actually, Einstein had made his atheistic views evident very early (1915) in his public life when he said: *"I see only with deep regret that God punishes so many of His children for their numerous stupidities, for which only He Himself can be held responsible; in my opinion, only His nonexistence could excuse Him."* Einstein once optimistically reflected on the future: *"The religion of the future will be a cosmic religion. It should transcend personal God and avoid dogma and theology. Covering both the natural and the spiritual, it should be based on a religious sense arising from the experience of all things natural and spiritual as a meaningful unity. Buddhism answers this description. If there is any religion that could cope with modern scientific needs it would be Buddhism."*

The next time someone tells you that Einstein really believed in God, please read to them what he actually said. Einstein's position is hardly unique. A reporter once asked our most famous contemporary theoretical physicist, Stephen Hawking, if his use of the phrase *"mind of God"* in his 1988 bestseller, *A Brief History of Time*, indicated that he believed in God. Hawking followed Einstein's lead and replied simply: *"I do not believe in a personal God."*

It is worth noting that both Einstein and Hawking were discreet in saying they did not believe in a "personal God." That doesn't mean they believed in an ultimate "Creator"—they understood that they could never disprove something for which there is no evidence. That is, as good scientists, they left the door slightly ajar for a possible creative force that we know nothing about.

Declaring oneself an absolute atheist can carry severe social penalties in many religious cultures, including those in much of this country. A 2015 Gallup poll showed that 40 percent of Americans would not vote for an atheist candidate for public office. Seven state constitutions have tests that effectively bar atheists from holding public office. Many Americans feel that atheists are immoral, and attempts to bar them from teaching in public schools are common. Most American atheists in public service are still "in the closet"—the most extreme discrimination is seen in Mormon and Southern Baptist states. Fred Edwords of the American Humanist Association asserted, "Americans still feel it's acceptable to discriminate against atheists in ways considered beyond the pale for other groups." Of course it goes without saying that Islamic cultures surpass all others in their anti-atheist discrimination—in 13 Islamic countries, atheists can be executed.

I suspect most of us intuitively understand why extremely religious people are compelled to discriminate against atheists—we will fully discuss this problem in later chapters. For the moment, suffice it to say atheist belief constitutes a direct contradiction and threat to worldviews based upon religion. People cling to their religion-based world-views because they provide comfort and solace, but mostly because they shelter them from a fear of death. Ironically, that often-crippling fear of death was mostly instilled by the same religion that subsequently promises them salvation and relief from their death-anxiety. Discussions of Ernest Becker's work in later sections will further clarify this relationship.

In answer to the question posed in our chapter title—*Atheist or Agnostic—Does it Really Matter?*—well, yes, it does matter! People react quite differently to the labels. Agnostics are not perceived to be nearly the threat of atheists. Calling oneself an agnostic does not alienate the religious and minimizes public discrimination. But perhaps just as important, it doesn't threaten the emotional security of those who depend on belief, especially in their declining years. It seems like an act of "Christian" kindness to identify oneself as agnostic!

Lest you think I am being too soft-hearted permit me to conclude this section with a quote from one of the most virulent atheists to ever set pen to paper—if you are a believer, he is blasphemous and vile—if you are a freethinker, you may find his remarks pretty amusing! That wonderfully slanderous atheist was H.L. Mencken, who among his many rants once said: *"God is the immemorial refuge of the incompetent, the helpless, the miserable. They find not only sanctuary in His arms, but also a kind of superiority, soothing to their macerated egos: He will set them above their betters."*

Even if you find the quote offensive, please read the last line thoughtfully and consider members of arrogant faiths you have known. Major religions often have appeal *because* they are irrationally contemptuous and exclusionist of non-members— they are able to *"set you above your betters."* That's not very "Christian," is it?

Why Are Humans Religious?

"Fear of things invisible is the natural seed of that which everyone in himself calleth religion." —Thomas Hobbes

"Religion is the sigh of the oppressed creature, the heart of a heartless world, and the soul of soulless conditions. It is the opium of the people." —Karl Marx

"Religion is something left over from the infancy of our intelligence; it will fade away as we adopt reason and science as our guidelines." —Bertrand Russell

Religions probably first appeared when hominins evolved brains capable of relatively complex thought. Proto-humans needed to develop changing strategies for coping with predators and finding food, and as their social groups expanded they needed to both assist and manipulate their fellows. Some of the first evidences of complex abstract thought are seen in human burial practices. Neanderthals, our ancient cousins of 180,000 to 30,000 years ago, sometimes buried their dead, presumably to protect the corpses from scavengers. They apparently did not adorn the bodies or include goods in the graves but we still infer that they had an awareness of death and had respect for the dead.[5] By 28,000 years ago our direct ancestors, the Cro-Magnon, were performing elaborate burials that included carefully dressing the deceased, decorating with flowers, and adding tools and art objects to the grave. Their actions suggest that they entertained belief in a spirit that survived death and passed on to an afterlife. Since that time, almost all human cultures have exhibited some form of ritual in connection with death and have promoted belief in some form of afterlife. A refined self-awareness and attendant fear of death were major factors in the development of the earliest human religions. In his Pulitzer-prize-winning book *Denial of Death* (Simon and Schuster, 1973), Ernest Becker argued that all religions developed as devices that permitted humans to deny the finality of death. Becker's thoughts on religion are discussed in somewhat greater detail below.

In addition to a newfound awareness of their personal mortality, early humans were confronted by unpredictable events in an often-frightening world. They must have wondered what caused the cycles of nature—the seasons, the storms, the floods, the droughts. They must have asked why they sometimes suffered from hunger, injuries and disease. They recalled that as infants their parents cared for and protected them when they were frightened or hurt. It is understandable that as adults they would wish for and invent parental figures to aid them in difficult times. When frightening events occurred they appealed to these imagined parental figures, which we would recognize as their gods.[6] These earliest humans had only a general understanding of causation, and if they observed that their supplications to their "gods" sometimes led to positive change, their belief would be reinforced. (We can't fault them for mistaking correlation for causation—most contemporary Americans, including many members of our media, still don't grasp that distinction.)

We would also observe that proto-humans, like us and our dogs, almost certainly experienced dreams during their sleep, and most likely they encountered deceased relatives within those dreams. They would quite naturally come to believe that their ancestors continued to exist in another place. Primitive conceptions of immortality and Heaven easily became part of their worldview.

Robert Wright asserts in his 2009 book, *The Evolution of God* (Little, Brown & Co.), that the early proliferation of religion was facilitated by the self-interest of the emerging clerical class. In his words: *"Whenever people sense the presence of a puzzling and momentous force, they want to believe there is a way to comprehend it. If you can convince them you're the key to comprehension, you can reach great stature."* The earliest clerics, whom we might call shamans, pretended to understand nature. No doubt most of them actually believed that they understood natural events better than their peers, and they sought to convince the others of their powers. They probably created narratives to document their powers, and some of the characters in their stories evolved into deities and demons. Since then, various clerics have continued and expanded that practice.

Most of the earliest hunter-gatherer societies were animistic and recognized multiple spirits and deities of varying levels of authority. It was more practical to assign different deities to different natural forces. If one god disappointed them, they could always count on the others. Individuals and tribes could have their own personal favorites. Furthermore, polytheism had the advantage of being easily adaptable to changing circumstances—new gods could be invented and old ones could be demoted or deleted as appropriate. Concomitant to the evolution of deities, each culture independently developed its own guidelines for behavior. In the earliest polytheistic cultures, norms of behavior were not linked to the supernatural belief system. Small hunter-gatherer bands of 20–50 individuals behaved in accord with the social norms expected of them by their family and friends. Everyone knew everyone else—social norms were simple and practical, and did not require supernatural enforcement. As we will discuss later, social norms of right and wrong were derived largely from the innate human capacity for empathy and fairness.

With the advent of agriculture, however, social units became larger and the inevitable division of labor presaged social castes. Increasingly complex and often caste-biased social mores required firm reinforcement and came to be associated with the wishes of certain deities. These rules or morals were at first maintained by oral tradition and remained adaptable, but with the advent of writing, adaptability was greatly reduced. The relative permanence of written documents tended to enshrine ideas into inflexible dogma. After just a few generations they became "sacred texts," and after a few more they were considered to be *"The Word of God."*

There is much written about the transition from polytheism to monotheism, and you were probably taught that monotheism represented some kind of cultural advance. We would argue that this is largely a Western perspective. While there are a combined 3.7 billion people in the world who claim adherence to the Abrahamic faiths, there are still more than a billion individuals who belong to various polytheistic faiths.[7] The question we should ask is why the monotheistic Abrahamic faiths prevailed in Western culture. There is no easy answer, especially in light of the logical paradox regarding the existence of evil known as the "Theodicy Problem."[8]

We do know that a shared belief in a single, powerful god is more conducive to strong, centralized tribal governance. The loosely affiliated small family units of hunter-gatherers managed quite well with various groups having their own assemblage of gods. But with the advent of agriculture and greater labor and class stratification, a single god helped to unify an increasingly complex society. The moralistic, omnipotent, personal mono-god of the Abrahamic tradition scrutinized details of every person's behavior, and fear of His Divine punishment compelled followers to honor the group's social contracts. Members of such groups could trust one another to exhibit both direct and indirect reciprocity. It has been suggested that one of the main reasons for the rapid spread of Islam was that their strong social contracts for reciprocity and trust enabled believers to develop extensive trading networks. Individuals could be reasonably certain that fellow believers whom they might never have met would still share their standards of fairness. Islam prospered and spread with the growth of the extensive Arab trading networks.

There is another reason for the advance of monotheism that is not commonly acknowledged but may be quite significant— all three Abrahamic faiths have, through most of their history, seen themselves as the exclusive source of truth. And few religions in history have proselytized more fiercely than have the Christians and Muslims. Centuries ago, with the emerging numerical superiority of Christianity and Islam, Jews came to avoid proselytism, quite properly fearing persecution by the other two numerically dominant faiths. We should note that Christianity and Islam prospered at the expense of Judaism partly because they claimed to be "selling a superior product"— subscribers were promised eternal life in a beautiful paradise. In contrast, Jewish sacred texts make little mention of life after death, and the ancient Jews did not envision *"Sheol"* as an especially pleasant place. Thus, the Jews did not make the idea of an afterlife central to their belief system and were thereby less successful in recruiting converts. Similarly, the older polytheistic faiths were generally religiously pluralistic and were not known to actively proselytize. Today, Hinduism, the world's largest polytheistic faith, is still pluralistic, and many alternative paths to enlightenment are acknowledged. It is likely that Christianity and Islam simply "out-recruited" Judaism and the older, more passive, polytheistic religions.

There are a number of features of human psychology that predispose us to embrace religion and believe in invisible deities. Foremost, humans possess an incredible capacity for indoctrination. This trait presumably evolved to facilitate rapid and effective socialization in our highly social species. The sooner individuals learned to observe the expectations and mores of their particular culture (i.e., "became socialized"), the more successful they would be. There is a definite survival value for children to quickly learn unquestioning obedience to their seemingly infallible, law-giving parents. Thus, many children are easily indoctrinated to belief in an infallible, law-giving, divine "Father."

Conformability is another related trait that facilitates our embrace of religion. Humans, like sheep, tend to obey leaders and follow one another—small wonder that religious leaders speak of their "flocks." Individuals conform in order to be accepted by their group—adopting the belief system of one's family and tribe is normally an adaptive trait. The human *tribal instinct* contains elements of both conformability and the human *need to belong to groups that afford social and/or physical security*. People are drawn to exclusionist groups that claim to be superior and are competitive with and/or militant toward alternative groups. Membership in primitive tribal warrior groups, fraternal organizations, soccer fan clubs and modern motorcycle gangs are exemplary, but the ladies' "Red Hat Society," the Elks Club, the Boy Scouts and most church memberships are equally valid examples. *Tribal membership* often gives one a sense of superiority (i.e., "self-esteem") whether or not it is deserved.

Religiosity is a complex sociological concept that generally means as it sounds—the general tendency to embrace belief in religion. There is considerable variation in the level of religiosity within populations, but the trait is known to be heritable and is therefore biologically determined to a certain extent. Some people are apparently born with particular gene configurations that predispose them to more easily accept things on faith and, accordingly, they tend to be more religious than others. Numerous research studies, including Tom Bouchard's famous work with "twins reared apart," estimate the heritability of religiosity at 0.5, or 50 percent. Please note that "religiosity" and "spirituality" are sometimes used interchangeably—that is incorrect on several levels, but most simply because spirituality does not require any belief in the supernatural.

In addition to these underlying psychological propensities there are also definite neurological mechanisms that facilitate belief. The emerging discipline of neurotheology, sometimes called "spiritual neuroscience," examines the relationship between neural mechanisms in the brain and the subjective experiences we describe as religious and/or spiritual. Neurotheologists assume that there are neural bases for religious experiences and that such mechanisms, just like the brain in which they reside, are the products of human evolution. Let us briefly review some of the more interesting and compelling research of this field.

Extreme religiosity commonly emerges in humans who suffer from neural aberrations. Some psychoses such as delusions and hallucinations have been associated with religiosity, but the causal relationships have not always been clear. Some writers have actually suggested that religiosity is a possible cause of schizophrenia, while others insist there is no causal relation between the two. However, we have known for some time that people suffering from epilepsy may exhibit increased religiosity. Many patients claim to "meet God" during their seizures and begin to act in religiously significant ways, including obsessive preaching or evangelizing. Sometimes their behavior exceeds normal bounds and they are described as "hyper-religious" or suffering from "toxic faith." In the extreme, such people can be a threat to themselves and others. More commonly, they may join cults, found new religions, or just become annoying acquaintances. Dartmouth University neurologist Gregory Holmes, in his studies of this phenomenon, examined the life of Ellen G. White (1827–1915), spiritual founder of the Seventh-day Adventist Church. Ms White was a prolific writer and had hundreds of religious visions that were central to the establishment of the Church. As a nine-year-old girl she had been struck in the head by a rock thrown by a schoolmate. Following the injury she lapsed into a coma that lasted for several weeks. During her recovery her personality abruptly changed—she became extremely moralistic and religious, and began seeing *"powerful religious images."* Holmes concluded that the blow to her head precipitated temporal lobe epilepsy (TLE), a localized form of epilepsy frequently associated with hyper-religiosity. (The 12 million living members of the Seventh-day Adventist Church might be unhappy with the suggestion that their founding matriarch was a delusional epileptic. Or perhaps not; it is widely believed that St. Paul was an epileptic who experienced his transformative seizure on the road to Damascus.) In recent surveys, four percent of all epilepsy patients report experiencing *"religious premonitory symptoms or auras"* prior to seizures, and about three percent of TLE patients have *"religious experiences"* in connection with their seizures. The incidence is significantly higher for those with right-side TLE (see note **6**).

Distinguished neuroscientist Michael Persinger of Laurentian University has been able to induce a profound "religious state" similar to that which occurs during TLE seizures using weak magnetic stimulation of the right temporal lobe. Many of his subjects experience a *"sense of timelessness and spiritual well-being,"* and some have claimed that they came *"face to face with God"* during the procedure. Persinger facetiously calls his experimental apparatus the "God Helmet"—fully 80 percent of his subjects report that they felt that they were *"not alone"* and that they sensed a *"spiritual presence"* when their right temporal lobe was appropriately stimulated. Of course, subjects tended to interpret their experiences in terms of their pre-existing belief system—thus far, no confirmed atheists have "met God" in the experimental program!

Closely related to this phenomenon is the so-called "third man factor" or "guardian angel effect"—the peculiar sensation that some people experience during periods of extreme danger and/or stress. Numerous soldiers, adventurers, daredevils and explorers have claimed to sense a comforting invisible presence much like that described in the fourth line of the 23rd Psalm (*"Yea though I walk* ... etc.). Persinger points out the similarity to temporal lobe stimulation and suggests it may be the result of vestigial "bicameral" activity in the stressed human brain (you may want to revisit footnote **6** for a better understanding). The "third man" effect and similar hallucinatory experiences predispose us to embrace the idea of supernatural supreme beings.

Consider next the nature of religious love/rapture—many a child, myself included, has experienced a powerful sense of awe during moving religious services. We were taught to love God and Jesus just as Muslims are taught to love Allah and Mohammed. Most religious people learn to experience a love of their deities and prophets. When we analyze the neurological and endocrine bases of these feelings of love for the Divine, we discover that they are physiologically identical to the mechanisms of interpersonal love. Sensations of both interpersonal and religious love involve the same brain loci, the same neurotransmitters (e.g., dopamine and norepinephrine), and the same hormones (e.g., oxytocin). We know our pre-human ancestors pair-bonded and experienced parental love long before they had religions. This leads us to suspect that religious awe/love is simply "parasitic" upon and evolved from the pre-existing, adaptive neural mechanisms of interpersonal love.

While I have real reservations about some of the supporting research, I am obliged to comment on the role of the peptide hormone oxytocin because it has received so much attention in the media lately. Oxytocin has long been known to play a central role in childbirth, lactation and mother-infant bonding in humans, but recently there has been interest in its apparent role in other social behaviors. It has been variously called the "love hormone," the "God hormone" or the "moral molecule." (The title of Paul Zak's 2013 book, which I will *not* make time to read, *The Moral Molecule—The Source of Love and Prosperity*, is transparently avaricious—or more precisely, *immoral!*) When you get past all the hype on oxytocin you discover there is some evidence that it does have effects on human prosocial and affiliative behavior, most especially social trust. Higher levels of oxytocin correlate with greater levels of trust (as well as trustworthiness) by individuals during interpersonal transactions. It appears that elevated oxytocin levels may enhance the individual's ability to recognize facial expressions and emotions and thereby understand the needs of others. The resultant enhanced levels of empathy and trust would reinforce reciprocal altruism and facilitate positive social and economic transactions.

Oxytocin levels are increased during a great variety of awe-inspiring events (entering a great cathedral or a redwood forest, attending a thrilling sporting event or a moving religious ritual, experiencing Niagara Falls or the Grand Canyon for the first time, hearing a stirring speech or sermon, experiencing orgasm, etc.). Accordingly, some writers have attempted to directly link oxytocin to increased religiosity and spirituality. Zak and others assert that individuals with higher oxytocin levels are happier, more secure and more self-confident. The positive effects of oxytocin are too often exaggerated while the negative effects are ignored. Short-term treatment with oxytocin may reduce symptoms of depression, but the long-term effects of the drug are quite different and potentially dangerous—the "love drug" appears to intensify negative emotional memories leading to increased anxiety and stress long after a stressful event has transpired. The effects of oxytocin on human behavior are far too complex to describe in this brief discussion. Suffice it to say, it should not be called the "love hormone" and should not be used psychotherapeutically until it is been more completely studied. I can find no clearly defined, consistently positive connection between oxytocin and religiosity.

As our earliest human ancestors entered new physical environments and social circumstances it was to their advantage to assess potential threats—they developed a cognitive system that enabled them to detect agency (intentionality) in others. Some researchers believe that their "agency detector" was so useful that it became overused and over-active—and that a misdirected and "hyperactive agency detector device" (HADD) led humans to imagine intentionality in non-human entities. In fact, imputing intentionality to unknown or inanimate objects could have survival value. For example, it was safer to assume that the large black object was a bear with bad intentions rather than a rock. Momentarily mistaking a rock for a bear had few long-term consequences, but the reverse could be fatal. Better to impute agency (intention) when in doubt—natural selection probably favored the HADD. Thus, it was adaptive for early humans to assume that supernatural agencies prevailed in the otherwise natural world. Those agencies would give rise to superstitions and the belief in spirits, devils, angels, and gods that characterized our first animistic and polytheistic religions. Our modern religions, with all their extensive supernatural trappings, may be viewed as the unfortunate byproducts of the once-adaptive ancient mechanisms. Space does not permit me to fully detail all aspects of this research, but suffice it to say that many cognitive scientists today consider religiosity to be, in large part, the coincidental byproduct of hyperactive agency detection devices in the human mind/brain. The curious reader should consult Shermer (2011, see below) for an excellent discussion of "agenticity" and "patternicity"— terms he uses to describe the innate human tendency to perceive agency and patterns, even when they aren't there.

What can we conclude regarding the human affinity for religion? First, humans have invented religions to mitigate their fears, most especially their fear of death. In our ancient past we became aware of the tenuousness of our existence and constructed a reality system that sheltered us from our fears—believers were spared the constant fear of death. Little has changed in the past ten thousand years—the main function of modern religion is to provide followers with a sense of security and hope for a better future. These seem like noble purposes, and religion thus appears to be advantageous to humanity. However, it is not clear that religion itself enhances biological (evolutionary) fitness and may simply be "parasitic" upon the older adaptive neural processes from which it arose. We will say more about the adaptive and maladaptive features of religion later.

Second, we can conclude that humans possess a variety of innate neural mechanisms that predispose us to "sense" supernatural entities. Our human sense of awe at spectacular events was conveniently appropriated to support our delight with supernatural inventions. Various brain pathways, neurotransmitters, and hormones that evolved to reward natural pleasures have come to also reward imaginary pleasures. And we know that the schizophrenic, epileptic or stressed normal brain are all capable of generating delusions that are variously called "the Third Man," a "Guardian Angel," or "God." It is likely that every human brain has a primitive "bicameral" ability to hear the "voice of God" in appropriate circumstances.

Throughout history, the majority of human beings have responded to these ancient neural mechanisms and have come to embrace religion and a belief in a supreme being. And yet throughout history there has also been a small but significant minority, usually from the more learned classes, that did not accept the supernatural explanations and who often viewed religion as an impediment to human progress. It appears that as humanity became more enlightened, the proportion of nonbelievers increased. Many of these dissenters came to see religion as a negative force in human welfare. We will consider this conflict in more detail below. In the meantime, let me conclude this section with a quote from a most prominent and controversial American theologian, John Shelby Spong, retired bishop of the American Episcopal Church.

> *Religion is primarily a search for security and not a search for truth. Religion is what we so often use to bank the fires of our anxiety. That is why religion tends toward becoming excessive, neurotic, controlling and even evil. That is why a religious government is always a cruel government. People need to*

understand that questioning and doubting are healthy, human activities to be encouraged, not to be feared. Certainty is a vice not a virtue. Insecurity is something to be grasped and treasured. A true and healthy religious system will encourage each of these activities. A sick and fearful religious system will seek to remove them.

The reader who is seriously interested in why we humans believe might want to read Michael Shermer's 2011 book, *The Believing Brain*. Shermer is an award-winning author, psychologist and science historian who has spent his professional life studying how and why people believe the things they do. In this very readable and complete book, he describes in simple terms the neural mechanisms that enable and often predispose the human brain to accept irrational beliefs and ultimately embrace belief in god, soul and afterlife.

Others of you may wish to just read Douglas Adams' take on belief in *Dirk Gently's Holistic Detective Agency*:

High on a rocky promontory sat an Electric Monk on a bored horse. ...

Electric Monks believed things for you, thus saving you what was becoming an increasingly onerous task, that of believing all the things the world expected you to believe.

Unfortunately this Electric Monk had developed a fault, and had started to believe all kinds of things, more or less at random. It was even beginning to believe things they'd have difficulty believing in Salt Lake City. It had never heard of Salt Lake City, of course.

The Roots of Human Morality

"Secular schools can never be tolerated because such schools have no religious instruction, and a general moral instruction without a religious foundation is built on air; consequently, all character training must be derived from faith ... we need believing people."—Adolph Hitler

"Freedom prospers when religion is vibrant and the rule of law under God is acknowledged." —Ronald Reagan

"I don't know that atheists should be considered as citizens, nor should they be considered patriots. This is one nation under God." —George H. W. Bush

"When I do good, I feel good. When I do bad, I feel bad. That's my religion."—Abraham Lincoln

"Ethical morality is doing what is right, regardless of what you are told. Religious morality is doing what you are told, regardless of what is right." —Anonymous

(For your enlightenment I've offered five different quotes above—I consider the first to be the religiously perverted ravings of a madman and the next two products of severely limited intellect. You may chose whichever you like best, but I am compelled to point out several facts often dismissed by revisionist historians. First, everyone should remember that Adolph Hitler was raised and confirmed in the Catholic Church. As late as 1941 he was quoted as saying, *"I am now as before a Catholic and will always remain so."* Of course few today would consider him to be a Catholic, but the important point is that he thought of himself as such. Biographer John Toland tells us that Hitler often proclaimed the Christian roots of German culture that enabled him to rationalize his most cruel acts. Toland said Hitler *"carried within him its teaching that the Jew was the killer of God. The extermination, therefore, could be done without a twinge of conscience since he was merely acting as the avenging hand of God."* Second, Hitler detested atheists and communists, and often linked the two by referring to "atheistic socialism" in much the same way American conservative Christians today speak of "godless communism." Third, Hitler expressed a hatred for organized labor unions—in part because they seemed socialistic, but principally because they constituted a threat to the power of the state. Upon assuming control, the Third Reich immediately sought to eliminate trade unions and any organizations that espoused or supported atheism.

Now consider "The Gipper," aka "Teflon Ron" Reagan. His famous "Reagan Doctrine" was in part predicated on his belief that "no one who disbelieves in God and in an afterlife can possibly be trusted." Although I don't think of Reagan as an evil man, I must reluctantly point out that he shared a number of interesting views with Hitler—he believed in Christian superiority and that he was doing God's work; he thought that atheism was evil and that "godless communism" was the most serious threat to human welfare; and, of course, like Hitler, he saw labor unions as potential threats to the stability of the nation. It is worth noting that like Hitler, Reagan was most popular with conservative Christians seeking a restoration of "traditional values." And of course, like Hitler, Reagan secretly consulted astrologers on national policy decisions.

I won't bother to consider the worldview of George Herbert Walker Bush because I doubt he ever had one of his own. I would note that one of his most famous policy statements was known as his "New World Order" speech, the exact same title as Adolph Hitler's equally famous speech of fifty years earlier—at the very least, an unfortunate (stupid?) choice of title. Bush's quote above reveals his pathetic ignorance of our patriot founding fathers, most of whom were deists or agnostics (e.g., Franklin, Jefferson, Madison, Washington, Paine).[9] Furthermore, he apparently didn't know that the words "under God" were added to the Pledge of Allegiance in 1954 after intensive lobbying by the Knights of Columbus in their attempt to shield us from "Godless communism" during the ruthless McCarthy anti-communist witch-hunts. Bush's selection of Dan Quayle as his VP running mate and his ignorant and insensitive comments should tell you everything you need to know about him.[10]

In my early teen years I read about the life of Christ and took to studying the New Testament on my own. I discovered the poetic beauty of the King James Bible and I "lost" the Douay Bible (along with the rosary beads) that my Aunt Margaret had given me. The more I read, the more I became convinced that Christ's divinity was irrelevant to his central message. My perspective was somewhat more cosmopolitan than my peers and siblings in that I acknowledged that the majority of the world's population was not Christian, let alone Catholic. I considered the fate of the untold numbers who walked the earth before the Christian era, and understood that one's religion was almost invariably a coincidence of birth. I embraced the study of science and sought rational explanations for the mysteries of the universe. When I asked myself the purpose of the contemporary divinity thesis, the answer was clear—control and power. If you can convince people that an invisible supreme being determines their ultimate destiny and that only the initiated will be saved, then you control them. Moreover, it became apparent to me that anyone who required either the promise of eternal life or the threat of eternal damnation in order to lead a decent and charitable life was not much of a "Christian." Einstein said the same thing more elegantly: *"Man would indeed be in a poor way if he had to be restrained by fear of punishment and hope of reward after death."*

Contemporary Christians, like Hitler, Reagan and Bush (quoted above), often argue that people must embrace religion because it is the source of all morality, and without religion people are amoral. Of course, that's ridiculous. Think about it for a minute. I'm an atheist and you are a good Christian; does that mean I must want to rob and murder you? Suppose you lost your faith and became an atheist—would you suddenly turn on me and rob and murder me? I don't think so.[11] Neurophysiologists and psychologists have monitored the emotional brain activity and behavioral choices of experimental subjects shown scenes of situations requiring moral action (*e.g.,* a child threatened with grave danger). Atheists, Muslims, Jews, and Hindus all showed the exact same neural patterns as fundamentalist Christians. They all exhibited emotional alarm and sought to act "morally" (*e.g.*, save the threatened child). They acted just as if they were all similarly "hard-wired" to act in a moral fashion—that's because they are!

The human capacity for moral action is pan-cultural and is found among Christians, Muslims, Jews, Buddhists, Hindus, atheists and agnostics, and apparently was present in our prehistoric ancestors as well. It is clear that human morality predates organized religion by tens of thousands of years. We are all born with an innate capacity for morality (a "moral compass") and come equipped with mechanisms ("moral emotions") to reinforce appropriate behaviors. Every normal human brain has a capacity for feeling *pride* and *guilt*—that is, our brain is designed by evolution to reward us when we do good and punish us when we don't—that's the essence of the Lincoln quote above. *Sympathy* prompts us to act altruistically toward others, and we feel *gratitude* when others act altruistically toward us. *Indignation* is elicited when we see others fail to appropriately reciprocate altruistic acts. You don't learn pride, guilt, sympathy, gratitude, or indignation *per se*—you only learn the mores of your particular culture. Your brain will automatically apply the appropriate emotional reinforcers when called upon. It should be clear that these same moral emotions serve to promote and reinforce cooperation and reciprocal altruism, and that these "moral" behaviors are rooted in the human genome. You are designed by evolution to be "good," even if your culture or religion corrupts that innate goodness. A number of modern writers convincingly argue that organized religions actually impede human moral progress. (The interested reader should consider reading Sam Harris's *The Moral Landscape* [2010] and Michael Shermer's *The Moral Arc* [2015].)

Of course "religious good" is largely dependent upon culture. Many religions condemn infidels and teach that killing them is good. You can find that hateful rhetoric in the Hebrew Bible, the New Testament and the Koran—each condemning the others to hell. If you are a "good Christian" your belief in the divinity of Jesus is absolute. Yet the Koran clearly states that you will go to hell for eternity for believing that (*"Allah has, surely, cursed the disbelievers, and has prepared for them a blazing fire, wherein they will abide forever,"* Sura 33:65–66). That is not a minor misunderstanding—it is an irreconcilable difference. It is one of many cases where a more recent religion condemns believers of an older established religion in order to usurp their allegiance. Muhammad did not restrict his hateful rhetoric to Christians—elsewhere (Sura 3:60) he asserts that Allah cursed the Jews and *"some were turned into apes and swine"* for turning *"farthest away from the Right Way."* In the *Hadith* of Bukhârî (Volume 1, Book 8:42) he is seen to condemn Christians and Jews together (*"May Allah curse the Jews and Christians for they built the places of worship at the graves of their prophets."*). You may want to think that those are uncommon and obsolete beliefs found only among the uneducated Muslims in Asia and the Middle East. Unfortunately, recent surveys of young, second-generation Muslim men in the US and the UK reveal they are often far more traditional than we think—roughly one-third think that suicide bombings are often justified and that individuals abandoning Islam (apostates) should be killed.

Early Christians condemned Judaism in much the same way that Muslims condemned Christians. According to St. John (8:44), Jesus was speaking to the Jews about their god when he said, *"Ye are of your father, the devil, and the lusts of your father ye will do. He was a murderer from the beginning, and abode not in the truth, because there is no truth in him."* St. Matthew (23:31) was more explicit in condemning the Jews: *"You snakes, you brood of vipers! How can you escape being sentenced to hell?"*

When Moses descended Mt. Sinai carrying the Ten Commandments and effectively founded the first of the three Abrahamic faiths, he emphasized that Jehovah would tolerate no other gods (Exodus 20:3), thereby condemning all earlier religions. And don't for a minute believe that the "great" Abrahamic traditions teach tolerance! Their intolerance and hatred for one another is not just an ancient animosity, as some would have you believe. In 2006, Pope Benedict XVI *"inadvertently"* quoted a 14th-century Byzantine emperor: *"Show me just what Muhammad brought that was new and there you will find things only evil and inhuman, such as his command to spread by the sword the faith he preached.*" As recently as April 2010, Franklin Graham, son of "America's pastor," Billy Graham, denounced Islam as *"an evil and wicked religion"* and said Muslims and Hindus don't pray to the same god as he. And we saw a spectacular example of vicious internecine Christian warfare in 1988 when the Reverend Ian Paisley, ordained minister of the Free Presbyterian Church and Minister of Parliament, publicly denounced the Pope as the Antichrist whom he described as *"a liar ... the original liar from the beginning ... he will imitate Christ ... Satan transformed into an angel of light, which will deceive the world."* Contemporary intranecine strife was evident when Rome-based Catholic Archbishop Raymond Burke charged that Boston Cardinal Sean O'Malley was acting under the influence of Satan, *"the father of lies,"* when he agreed to preside over the funeral of pro-choice Catholic Teddy Kennedy. These are just a few of my favorite examples—there are thousands more!

To Rodney King's familiar plea—*"Can't we all get along"*—I must respond "Not a chance in Hell, Rodney, but thanks for the suggestion!" These are not moral people (*"except perhaps you, Rodney"*), so please don't suggest to me that religion is the root of morality. Their vicious sectarian rhetoric corrupts all who follow them. I should note that recent research clearly demonstrates that extreme sectarianism (*i.e.*, certainty that your religion is the one true faith) is found almost exclusively among individuals whose reasoning and information processing skills (IQs) are significantly below average. This recalls Albert Ellis's observation cited previously: *"The only thing we are not sure about is whether only stupid people become religious, or whether religion makes them stupid."* I suspect it is a little bit of both!

Personally, I'm content to take my morality not from some hateful religion but rather from what I know to be evolutionarily based "human nature." Contrary to that ridiculous doctrine of "Original Sin," humans, with rare exceptions, are born with innate goodness. We have the inborn tendency to act cooperatively and altruistically because our ancestors who acted cooperatively thrived, and those who did not, often failed. That's simple natural selection. Explaining altruism is a little more complicated, but we know that we come equipped with those "moral emotions" mentioned earlier, and that they also reinforce reciprocal altruism. That is, if I do you a favor, you normally reciprocate at a later date. If you fail to reciprocate, you may experience guilt and I may exhibit indignation and anger. My friends will sympathize with me and share my indignation, and you will likely be censured by our group. If they identify you as a "cheater" they may withhold aid to you at a later date.

I could go on at great length, but suffice it to say that there is now an enormous body of real research data by sociobiologists, evolutionary psychologists, anthropologists, political scientists and economists that elucidates the evolution and development of human reciprocity, altruism, cooperation and sense of fairness.[12] Humankind is not innately selfish or sinful, as some have posited—while we all have an instinctive capacity for self-survival, the exaggerated form we know as "greed" is largely learned and is often maladaptive. Thus, we may safely conclude that "goodness" has very little to do with religion. Humankind comes equipped with an innate sense that enables us to act ethically without religious coercion or a belief in the supernatural. (I will not waste my time and yours explaining why Ayn Rand's objectivist philosophy and rejection of the *"ethic of altruism"* is so much bullshit. Simply stated, her philosophy is based on a pathetically sophomoric grasp of human biology and evolution.)

Much that has traditionally been considered "religious morality" is really "human nature." The "Golden Rule," for example, is embraced by virtually every religion from ancient Egypt to modern times. They all claim to have invented it but, of course, none of them did. The "Golden Rule" of Judeo-Christian tradition is merely innate reciprocal altruism expressed as dogma. It did not originate as a religious teaching; it is a genetically based universal human quality. All humans born into normal cultures will develop this sense of fairness unless taught otherwise.[13] There is evidence that it existed in prehistoric human (Cro-Magnon) cultures and it is found in contemporary polytheistic/animistic societies.

Recent studies have identified specific brain centers associated with moral judgments. In order to make such judgments we need to infer the mental state of others, including their intentions, desires, beliefs, etc. This ability, known as "theory of mind" (ToM), enables humans to sympathize, empathize and act fairly. The precise brain mechanisms are currently being elucidated.[14] Modern neuroimaging methods have linked these physiological and behavioral attributes to specific brain regions—for example, ToM function is associated with a brain center near the right temporoparietal junction, and the moral indignation associated with unfairness is expressed via a plum-sized area in the ventromedial prefrontal cortex.

Overwhelming evidence now suggests that "human morality" is the product of natural and ultimately decipherable biological processes that prevail during normal development. Negative environmental and/or cultural factors can sidetrack the normal process, but the vast majority of the world's people are humane and charitable individuals despite their lack of a "Christian education." In fact, when religions indoctrinate with their god-fearing dogma, their followers often pervert their innate moral compasses in the service of that dogma. While some religious individuals may believe that atheists are less moral, the available data show the opposite—atheists, on average, are more moral, ethical and generous than most "believers."[15]

THE PRINCIPAL EVILS OF ORGANIZED RELIGION

"Those who can make you believe absurdities can make you commit atrocities." —*Voltaire*

"Men never do evil so completely and cheerfully as when they do it out of religious conviction." —*Blaise Pascal*

"My own view on religion is that of Lucretius. I regard it as a disease born of fear and a source of untold misery to the human race."—*Bertrand Russell*

"Religion teaches people to be extremely self-centered and conceited. It assures them that god cares for them individually, and it claims that the cosmos was created with them specifically in mind."—*Christopher Hitchens*

Organized religions usually spring from some sort of preposterous claim of Divine revelation. We are told that around 1400 BC God spoke to Moses on Mt. Horeb (Sinai) and gave him two stone tablets bearing a list of ten moral imperatives. We are also told that three millennia later (AD 1827) on "Hill Cumorah" in Manchester, NY (not far from where I-90 today intersects NY Route 21) Joseph Smith, Jr. received from God's agent ancient golden tablets bearing a sacred text. (The angel was a Native American with the improbable name of Moroni— shall we call it a *moronic* revelation?) Thus were the Ten Commandments and the Book of Mormon delivered to save humankind. (Incredibly, despite their crucial importance to our salvation, we seem to have misplaced both sets of sacred tablets. Damn!) Most major religions claim Divine origins of their sacred texts. In just a few generations such Divine revelations can become dogma, and we have a new religion. It can happen quickly—sacred information received by L. Ron Hubbard's *"thetan"* (soul) during a previous extraterrestrial life became part of the *"Upper Levels"* of the Church of Scientology as recently as 1967. Today that sacred knowledge is revealed only to the most devoted (and generous) members of the Church (e.g., Tom Cruise and John Travolta). The "cargo cults" that sprang up among Pacific Islanders during WWII became established religions in just a few years. The Jon Frum cult has a wide following in Vanuatu to this day.[16]

Am I being blasphemous by discussing Moses, Joseph Smith, that obvious fraud L. Ron Hubbard, and the prophet John Frum in the same paragraph? I certainly hope so, because I see little real distinction between them and their accomplishments. In all four cases religious leaders emerged and exploited the weak and gullible. In return the followers were made to feel superior and privileged and were, in the first three cases at least, taught that they might earn some form of eternal reward for their faithfulness. Such obedient followers often attain a level of pride and arrogance not justified by their earthly moral accomplishments.

Individual religions, usually early in their development, define the special rituals and sacrifices that their followers are expected to practice to prove their devotion and worthiness. These may include forms of prayer, ritual sacrifice, self-mutilation, excessive gifting (e.g., "tithing"), and such. In reciprocation for these "sacrifices," the religion makes promises to its followers. Eternal life is one of the most common—it may be as reincarnation or eternal life in a place called Heaven or Paradise.[17] Of course the idea of an afterlife has enormous appeal and its acquisition often becomes the sole objective in many Christian and Muslim lives. Adherents are told that if they live the religiously obedient life on Earth, they will live forever in Heaven. We humans have great difficulty confronting our own mortality because evolution did not design our brain (mind) to easily comprehend our own non-existence—thus, it is natural that immortality would become the principal default perspective for humans. As increasingly self-aware humans were forced to consider the prospect of our own deaths, cultures developed life-after-death myths to eliminate the otherwise confusing and frightening passage to non-life. The invention of the "soul" provided us with the necessary vehicle, and our religions provided us with the maps and signposts for the fanciful trip to Paradise.

It is an unfortunate fact that most new religions quickly come to substitute religious sacrifice for acts of goodness. If the dutiful practice of appropriate rituals becomes sufficient to ensure safe passage to the hereafter, then acts of goodness to others become irrelevant. We can understand why the many millions of people worldwide who suffer indignities of subservience, poverty and ill health come to embrace religious ritual. Their belief in a better life after death is a major sustaining force in their often short and unpleasant lives. But that belief which sustains them also holds them in intellectual, emotional, religious and economic bondage—if they quietly and obediently go about their miserable lives, they are assured safe passage by their spiritual leaders. Karl Marx saw religion as a tool of the ruling class—the masses were relieved of their fear and pain through religious emotional experiences and comforted by their knowledge that suffering in this life would lead to eventual happiness. Napoleon Bonaparte may have expressed it best when he said, *"Religion is what keeps the poor from murdering the rich."* The ruling class knew that as long as the masses believed in religion, they would make no real effort to discover and overcome the true cause of their suffering. Some have argued that the world's religious leaders are often in collusion with the political leaders and that the clergy, by virtue of their hypocrisy, are the more guilty.

The following sections identify what I consider to be some of the major failings of organized religion. The categories are not mutually exclusive, and there is much more I could add to each. But in accord with religious tradition and the model of the Commandments, I kept the number of sections at ten! My comments are directed principally to the main Abrahamic faiths (Judaism, Christianity and Islam) but in many cases they apply equally well to others. And I must emphasize that my remarks are addressed to the religious organizations and hierarchies. I bear no animosity toward most religious people—some, not many to be sure, but some of my friends are religious.

I. Suppression of Scientific Progress

"The Church says the earth is flat, but I know it is round, for I have seen the shadow on the moon, and I have more faith in a shadow than in the Church." —Ferdinand Magellan

"Those who assert that the earth moves and turns are motivated by a spirit of bitterness, contradiction, and faultfinding; possessed by the Devil, they aim to pervert the order of nature." —John Calvin

"If science disappeared from human memory, we would soon be living in caves again. If theology disappeared from human memory, no one would notice." —Terry Sanderson

"Reconciliation between science and Christianity would mean squaring physics, chemistry, biology, and a basic probabilistic reasoning with a raft of patently ridiculous, Iron Age convictions." —Sam Harris

Since 1986 the Catholic Diocese of Buffalo has been working to secure sainthood for a most remarkable man, the late Father Nelson Baker (1841–1936). His many humane accomplishments included founding a home for unwed mothers and their infants, a boys' orphanage, a hospital, and several schools. The process of canonization normally requires evidence of two miracles, and at this point the Diocese has submitted only one to the Vatican. Not surprisingly, the "rules of evidence" for the Church are somewhat different than mine, and, I expect, yours. The Vatican's "validation" of a miracle is based upon a logical fallacy known as *argumentum ad ignorantiam* or literally, "appeal to ignorance." That is, in the absence of any knowledge to the contrary, a claim can be accepted as true. If the diocese claims an event in Fr. Baker's life was a miracle and there is no contradictory evidence, it can be considered "valid." According to rumors in the media, Fr. Baker's pending miracle involved a surprising recovery by a woman who had been declared terminal by her doctors. Shortly after relics from the Father Baker collection were applied to her legs, she experienced a complete recovery. The Church's reasoning is that "we can't explain it; therefore it must be a miracle." It is *argumentum ad ignorantiam*—it is not merely unscientific, it is pathetically *anti*-scientific. It reinforces the irrationality that is so widespread among the very religious today. It teaches that in the absence of compelling evidence to the contrary, we can accept anything we chose as truth. From most of what I've read, Father Baker was a extraordinarily compassionate and generous priest who performed something like "minor miracles" in his everyday service to the people of Buffalo. The Church should be satisfied to honor him for his very real work and not attempt to exaggerate his actions into something Divine and supernatural, which every rational person knows they were not. This is just one small example in the unending conflict between reason and religion.

Western religions have unavoidably been at war with science and reason for the past two millennia. Those who hope for reconciliation between science and religion are hopeless dreamers. As long as religions insist that they can know the truths of the universe through irrational, supernatural belief, the two can never reconcile. Religions almost always compromise rational truth—they require their followers to accept supernatural explanations for many obviously natural phenomena. Furthermore, they traditionally reward ignorance and punish enlightenment.[18] They are compelled to continually challenge and scrutinize scientific advances to determine if they are compatible with their various ancient *"sacred texts."*. Unfortunately, most of their sacred texts are nothing more than plagiarized ancient myths and civil codes. The findings of modern science are rarely compatible with the laws and legends of nomadic goat-herders who lived 5,000 years ago. Space does not permit me to discuss all the areas of conflict, so I will consider just a few major ones.

Fundamentalist Jews, Christians and Muslims today are unable to reconcile Darwinian and organic evolution with the Divine origins described in Genesis. They are obliged to teach that evolution is false and thereby deny themselves and their children an understanding of one of science's most fruitful and beautiful theories. To its credit, the Catholic Church recently acknowledged the validity of evolution as an explanation of plant and animal diversity, but regrettably, it steadfastly refuses to consider the human brain (mind) to be the product of evolution. Benedict XVI, the most recent past pope, had such conservative views that he threatened to return the Church to the 17th century. In his installation as Pope in 2005, he said, *"We are not some casual and meaningless product of evolution. Each of us is the result of a thought of God. Each of us is willed, each of us is loved, each of us is necessary."* That may sound profound and comforting to you, but it is pathetic theological pandering. The Catholic Church will never be able to understand evolution as long as it insists that man is a divine and distinct creation uniquely equipped with that vehicle of immortality, the soul.

Little has changed since the Church began its persecution of Galileo for advocating the heliocentric universe 380 years ago.[19] Today we have "young-earth creationists" who, despite overwhelming evidence to the contrary, insist that the earth was created by God in just six days a mere 6,000 years ago. In Petersburg, KY we have a new "Creation Museum" that attempts to depict earth history in such a Biblical model. Each year, thousands of schoolchildren are exposed to presentations by pseudo-scientists preaching anti-evolution nonsense. Not far away, near Williamstown, KY, a $150 million theme park called "Ark Encounter" features a full-size replica of Noah's Ark with a nearby Tower of Babel. In Seattle, WA we find the "Discovery Institute" with its "Center for Science and Culture." It is a bit more subtle than the "Museum" and the "Ark," but still distributes ridiculous anti-Darwinian literature to fundamentalist preachers and pro-Creationist school boards. I assume that anyone reading this is intelligent enough to understand why the higher courts have repeatedly denied demands for inclusion of *"Creation Science"* in biology curricula—because it is religion, <u>not</u> science! I have little patience with those Americans who don't comprehend that distinction. I am not encouraged by the 2010 Angus Reid Public Opinion Poll that revealed that 68 percent of Britons and 61 percent of Canadians understand that *"human beings evolved from less advanced life forms over millions of years,"* whereas 47 percent of Americans actually believe *"God created human beings in their present form within the last 10,000 years."* That's pathetic!

A more recent (2014) National Science Foundation survey provides some insight into how Americans really think—48 percent of respondents said they thought it was true that *"human beings, as we know them today, developed from earlier species of animals,"* but when the same question was prefaced by *"according to the theory of evolution,"* 72 percent said they thought it was true. When asked about the Big Bang theory, 39 percent agreed that the *"universe began with a huge explosion,"* but 60 percent agreed when the statement was prefaced by *"according to astronomers."* How do we account for the disparity? The respondents' personal beliefs were significantly more "conservative" than what they knew to be accepted scientific facts. What possible cultural force would persuade some people to doubt the validity of established scientific facts? I don't have to say it, do I? In the same survey, 23 percent of the respondents believed that the sun revolved around the earth. I suspect those are the same poor souls who believed *"God created humans in their current form within the last 10,000 years"* in the 2010 survey. It would be funny if it weren't so tragic.

Scientific literacy in the United States is at the lowest level of any time in recent history, and organized religion is largely to blame. More and more Americans are home-schooling their children, not because the public schools are so bad, but because their preachers tell them they must shelter their children from the evils of "secularism" found there. That's exactly the same reason that Muslims send their kids to "madrasas" to memorize the Koran and learn to hate Western culture and science. Just like their fundamentalist Christian counterparts, Muslim children are taught that evolutionary theory is evil.

Similar perversions of science justify fundamentalist opposition to stem-cell research and abortion. These people believe that each human life is Divinely mandated and each contains an elusive feature, the soul, placed there by the Creator. There never has been and likely never will be any credible scientific evidence for such an entity, but that doesn't prevent the presumption of its existence from having enormous influence on public policy. The current fundamentalist perspective is that the process of "ensoulment" (*i.e.*, when the soul enters the "new" human) occurs very early, presumably at the time of conception (when the ovum is fertilized by the sperm). It is this event that gives the "new life" sacred status. Of course the presence of a soul sharply distinguishes human from non-human life and reinforces humankind's sense of biological superiority—it is why we grant infinitely more rights to a tiny cluster of undifferentiated cells growing parasitically in a first-trimester human uterus than we grant to adult chimpanzees, whales, and even our pet dogs and cats. Even after 12 weeks, that differentiating mass of human cells has less sentience than an ant you might step on in your driveway. We will avoid getting into the inane discussions of "when new life begins" (life doesn't begin *de novo* each generation—life is a continuum) or the specious argument that *"the new life begins at fertilization because that is when the new genetic individual forms"* (not so; much of the genetic reshuffling occurs in meiosis prior to fertilization—so it is just as reasonable to say that new life begins at the formation of each ovum or sperm cell). Similar pragmatic positions such as *"life begins when the heart starts beating"* ("heartbeat" means nothing— rhythmic contraction is an intrinsic property of cardiac muscle cells) or *"when brainwaves first appear"* (fetal brainwaves reflect the spontaneous discharge of growing nerve cells and do not signal anything comparable to sentience or cognition) are equally fallacious. It is impossible to scientifically identify when new life begins because, as we said above, it doesn't begin—it is a continuum. Arbitrarily specifying fertilization or implantation makes no more sense than saying life begins at ejaculation (as some ancients believed). Scriptural references offer no help. The time when the baby *"quickens in the womb"* (when fetal movements are first felt by the mother—normally between 14 and 20 weeks) identified the start of human life to certain ancient Jews. That, however, did not convey the full rights of personhood— the same ancient Jews apparently condoned abortion and permitted parents to kill a "disobedient child" without it being considered murder.[20] Medieval Islamic law ruled that the fetus received its soul, thus becoming fully human, at 120 days, and physicians were allowed to induce abortions up to that time. If you hope to find prohibitions to abortion in the Judeo-Christian Bible you will be disappointed. In fact, the only specific reference to abortion actually *mandates* it in the case of a married woman impregnated by a man other than her husband (Numbers 5:12–28). Isn't it interesting that there are no other references to abortion in Scripture written during a period of history when abortion was exceedingly common?[21] It seems to me that if it were such an abomination to God there would be some mention of it among his seemingly endless lists of laws and rules. Jesus had to be fully aware of the widespread practice of abortion in his culture, yet never mentioned it. Don't you find that just a little curious?

If you have followed my rant to this point you probably recognize that I can find no valid moral, ethical, or scientific bases for the attempts by religious fanatics to deny women the right to an abortion before 20 weeks, or to block the use of embryonic tissues (stem cells) in medical research. Most of their arguments are ultimately based upon a belief that reproductive rights are somehow sacred and should be controlled by the Church—or worse, that the State should preferentially enforce particular religious beliefs regarding reproduction. Many of these same religious extremists condone the murder and attempted murder of abortion providers and the bombing of abortion clinics. These "faith-based" attacks are no different than many of the "faith-based" bombings and murders committed by Muslim extremists.

We are supposed to be a nation of laws, and we have courts to settle these things. In 1973 the US Supreme Court ruled in *Roe vs. Wade* that a woman's right to have an abortion was protected by the US Constitution. That ruling meant that states were prohibited from banning abortions performed prior to the point of fetal viability (*i.e.,* that age when the fetus can survive without heroic medical assistance). *Roe* originally established viability at 24 weeks, but a subsequent decision in 1992 (*Casey v. Planned Parenthood*) shortened it to 22 weeks. I am comfortable with that—as a biologist I know for certain that no sentient creature is being destroyed by a first trimester abortion, and much good might come from using the tissues from aborted fetuses for medical research and therapy.

Clearly connected to the religious opposition to abortion is the opposition to birth control. The bizarre theological basis of this belief, the story of Onan, is discussed in more detail below in Section V. The *real* basis of the Christian and Muslim prohibitions on birth control is more political than theological. Both religions have actively proselytized and sought to increase their numbers throughout history. Today some of the planet's highest birth rates are found in fundamentalist Jewish, Christian and Muslim populations. The propaganda mills in some Muslim countries insist that birth control is a Western plot to limit their growth. The same people who reject birth control and deny the global population crisis can't understand why more than half of their young men and women can't find work.

In the US few people make the connection between population growth and unemployment even though they understand that the economy must grow if the nation is to prosper. Americans periodically struggle with unemployment crises but fail to realize that we must create 120,000 new jobs every month just to keep pace with our population growth. Without a scientifically based national population policy that limits both immigration and child tax incentives, we are doomed to an extended period of overpopulation and under-employment. The global consequences of overpopulation will be devastating. (Did you ever wonder why such a majority of the 9/11 highjackers were from relatively affluent Saudi families? Do you understand the proverb *"Idle hands are the devil's playground"*?)

* * *

Let me conclude with the latest ominous statistics regarding science education in the United States. Depending on the metrics employed, we rank between 21st and 29th in the world in science education—just 10 years ago we ranked 14th. Twenty-five years ago we were near the top. Most industrialized Asian and European nations are now superior to us—even Croatia, Hungary, Slovenia, Estonia and the Czech Republic manage to beat us out. Among the 36 most industrialized nations we rank 18th. Our math scores are even worse. This does not bode well for this country's future innovation in science and technology. But even if we import our scientists (as we import so many doctors and engineers) we will not address the very serious problem of scientific illiteracy among our voters. We will continue to make misguided decisions with respect to national science policy and will fall further behind. Scientifically illiterate voters will be prone to making bad decisions on important issues facing the planet.

There are a few signs of progress—the latest Gallup poll (March 2017) reveals that 68 percent of Americans now believe that global warming is real and is caused principally by human activities. Unfortunately, 55 percent of them *"don't worry a great deal"* about it, and 58 percent don't think it will *"pose a serious threat"* in their lifetime. While not as scientifically illiterate as poll respondents a decade ago, they still fail to grasp the seriousness of the problem. It is worth noting that people who exhibit a strong religious belief in an afterlife are inclined to ignore the status of the planet.

So how did such scientific illiteracy come about? Parents like to blame teachers, and teachers like to blame parents. Politicians blame school administrators, and school administrators blame politicians. Of course they are all to blame to varying degrees, but the root cause is the cultural milieu that disparages science and encourages people to pick and choose what they want to believe because of the egalitarian nonsense that *"everyone is entitled to their own opinion."* Perhaps, but as the late Senator Pat Moynihan pointed out, *"everyone is not entitled to their own facts."* At this point I should not need to explain the major force perverting the American cultural milieu!

II. Tolerating/Promoting Child Abuse

"Any system of religion that has anything in it that shocks the mind of a child, cannot be true." —*Thomas Paine*

"It's time to question the abuse of childhood innocence with superstitious ideas of hellfire and damnation."
—*Richard Dawkins*

Throughout history, religious leaders have wielded enormous power over many millions of people, and more often than not they have abused that power. The horrendous religious persecutions of the Crusades, the Inquisition, the Reformation, and the European and American Witch Trials are well known to students of history. But some of the worst forms of abuse are the acts against helpless children that continue to this day. While we are all somewhat aware of the various forms of physical and sexual abuse, we should also consider the forms of mental and emotional abuse many children suffer as a result of their religious indoctrination.

I won't review all the sordid history of pedophile priests throughout the world, but I will note that during the course of the recent revelations the Vatican has acted mainly to protect its name (and power) and has shown little authentic concern for the welfare of its thousands of victims. Award-winning columnists Brooks Egerton and Reese Dunklin of the *Dallas Morning Sun* reported in 2002 that two-thirds of sitting US bishops had been accused of moving pedophile priests to new assignments. In their words, *"It's not a few bad apples; it's the barrel."* Imagine that a public middle-school teacher in your hometown had sodomized 200 young boys—do you think the State Commissioner of Education would reassign him to a different school to protect him? Few of the many priestly perpetrators will ever be defrocked, let alone subjected to appropriate criminal prosecution.[22]

Child rape is not the exclusive province of the Catholics. In September 2007 a Utah jury found Warren Jeffs, head of a splinter Mormon sect, guilty of being an accomplice to child rape. Fourteen-year-old Ellisa Wall had been forced into a "spiritual marriage" with her 19-year-old cousin, Allen Steed. When she appealed to Jeffs, leader of her church, he told her that God had told him that if she did not consent to the sexual advances of her new husband she would lose her salvation. The choice was clear—have sex with her 19-year-old cousin or burn forever in hell! The prosecution for the State cited earlier case law (*State of Utah vs Chaney*) in which a 13-year-old girl was promised the reward of eternal life in heaven for having sex with an older man. Sexual coercion by either threats of hell or promises of heaven is genuinely evil and constitutes child abuse to any rational person.

Most enlightened people consider the forced marriage of children to be a violation of basic human rights. But scores of countries still have significant child bride problems. Not until May 2014 did courts in England and Wales finally make forced marriages illegal. Parents and/or clergy who force children into marriage against their will can be imprisoned for up to seven years. I would say it's about time.

Obviously it's not just Mormons and Catholics abusing children—in April of 2010 the BBC reported that a 12-year-old Yemeni bride died of internal bleeding three days following her marriage to an older man. Yemeni law remains ambiguous because their scriptures condone child marriage—sacred Muslim texts tell us that the Prophet Mohammad took "Aisha" as his second wife when she was just six years old and consummated the marriage when she was nine. In 2009 Yemeni lawmakers passed legislation setting the minimum age of marriage at 17, but the law was soon repealed as being un-Islamic when clerics insisted that the Prophet Muhammad's actions should serve as their model. For centuries various critics have labeled the Prophet Mohammed a pedophile. Pederasty is still practiced in certain Muslim cultures—it is commonplace among the Pashtun of Afghanistan, where it is known as *bacha bazi* (boy play). Wealthy Muslim men often "own" prepubescent boys—but when the boys reach maturity they must be abandoned because homosexuality carries the death penalty.

We should mention female genital mutilation (FGM) as another brutal form of child abuse. FGM is practiced on prepubescent girls, normally 4–8 years of age. During the procedure the girl's genitals are partly removed and/or mutilated ostensibly to inhibit her sexual sensation as an adult. It is practiced throughout North Africa, the Middle East and parts of Southeast Asia. It is especially prevalent in Somalia, Egypt, Eritrea, Sudan, and Ethiopia, where between 75 and 98 percent of women are subjected to the procedure. FGM is found principally in Muslim countries but is also practiced by Coptic Christians and Ethiopian Jews. There is little mention of FGM in scripture, and it is believed to predate Islam—nonetheless, practitioners generally believe they are observing a mandated religious ritual. The Supreme Council of Islamic Research has ruled that FGM has no basis in Islamic Law and several leading Islamic scholars have called for an end to the practice. But not all Muslims agree, and the practice persists. The World Health Organization estimates that worldwide, between 100 and 140 million girls and women have been subjected to some form of FGM, and each year three million girls in Africa are at risk for the procedure. It won't end until religious leaders actively condemn the practice.

Child abuse is not just sexual or physical. Just as abhorrent are the requirements in various faiths that believers refuse modern medical intervention when their children are suffering from treatable conditions. If adults decide that they will simply pray for deliverance from their own diseases and injuries, that is their choice—it's stupid, but it's their choice. However, when parents decide so on behalf of their infant children, they are guilty of child abuse and should be prosecuted. In what was an exceptional case, in February 2014, a Philadelphia couple was sentenced to three and a half to five years in prison for defying a 2011 court order to get medical care for their children. The earlier court order followed the death of their two-year-old son who died of treatable pneumonia. The couple, members of a Pentecostal sect, chose to pray over their dying son rather than seek medical care. Despite the 2011 court order and being sentenced to ten years' probation, the couple repeated their "profession of faith," and two years later allowed a second son to die of pneumonia! This time the judge did not excuse them from jail time in deference to their religious beliefs—the earlier court order had clearly stipulated that they must have annual medical check-ups for their children and seek medical attention whenever the children were ill—they failed to do either. The sentencing judge told them, *"You've killed two of your children. Not God. Not your church. Not religious devotion. You!"* The couple has *seven* surviving children who are now in supervised foster care and attending public schools.

This form of abuse will continue as long as we continue to respect religious beliefs that are contrary to our scientific (medical) knowledge. Every year, children of Seventh-day Adventists, Christian Scientists, members of the Church of the First Born, Followers of Christ, First Century Gospel (Pentecostal) practitioners, and various other religious sectarians are denied appropriate health care by their ignorant (but very religious!) parents. Some have suffered painful deaths from treatable childhood diseases—meningitis, bowel obstruction, appendicitis, and pneumonia—as their parents and church elders substituted prayer for appropriate medical care. Other children are denied needed transfusions or vaccines. When such children die or suffer needlessly their negligent (but very religious!) parents typically invoke a "freedom of religion" defense to avoid prosecution. The American Academy of Pediatrics has recently urged that parents and others who deny children necessary medical care on religious grounds be subject to civil or criminal action.

I prefer not to get into a discussion of the insanity dispensed by today's "anti-vaccine" contingent and the incredible risk of diphtheria, tetanus, polio and measles to which they subject their own and others' children. One example will suffice. In 2005 a 17-year-old girl from Indiana contracted a virulent form of the measles virus while on a church-sponsored "missionary" trip to Romania. Upon her return she managed to infect 34 other people, appropriately enough, at a church picnic. Thirty-one of the 35 unvaccinated and three of the 465 vaccinated picnic attendees contracted the disease. The majority of the victims were home-schooled children—current Indiana law requires youngsters attending public school to have measles vaccinations, but the state has been quite liberal in granting "religious exemptions."

Finally, we would note that some of the cruelest forms of child abuse are psychological. Religious indoctrination in the Abrahamic faiths has traditionally involved detailed descriptions of the horrors of hell. The "Lake of Fire" myths were contrived to terrify young people into a lifetime of submission. Many Catholics will attest to the cruel effectiveness of that indoctrination. Richard Dawkins tells of an American Catholic woman who experienced two tragic events as a seven-year-old. First she was molested by her parish priest, and a few months later her best school friend, a little Protestant girl, died unexpectedly. She said the priestly molestation's effect on her seven-year-old psyche was merely *"yucky,"* but the certain knowledge that her dear little non-Catholic friend would spend an eternity burning in hell horrified her beyond description. I can identify with that—after more than 70 years I can still vividly recall my mother telling me that my father would be condemned to hell because he was never baptized in her Catholic faith. Dawkins' correspondent persisted in her conflicted Catholicism, whereas I was able to reject such cruel myths at an early age.

Dawkins made the following statement that has outraged many Catholics: *"Regarding the accusations of sexual abuse of children by Catholic priests, deplorable and disgusting as those abuses are, they are not so harmful to the children as the grievous mental harm in bringing up the child Catholic in the first place. Being taught about hell—being taught that if you sin you will go to everlasting damnation, and really believing that—is going to be a harder piece of child abuse than the comparatively mild sexual abuse."* You may not agree with all he says, but it should make you think. Do you think sexual abuse is worse than emotional abuse? Is your opinion the product of your Christian indoctrination? In the April/May 2010 issue of *Free Inquiry*, Chris Edwards raises the question, "Is Hell Illegal?" and argues that intimidating a child with hell is hardly different than threatening the child with any other form of violence. A father who threatened to shoot his daughter for not cleaning her room would likely end up in jail, but if he threatened her with eternal hellfire for the same offense, many "good Christians" wouldn't think ill of him. Distinguished psychologist Nicholas Humphrey has argued that we should free the children of the world from the religions that damage their minds when they are still too young to understand what is happening to them. As Dawkins says: *"Priestly groping of child bodies is disgusting. But it may be less harmful in the long run than priestly subversion of child minds."* In fairness, I would suggest that fundamentalist Islam and Judaism probably damage at least as many young minds as Catholicism.

III. Employing Dishonest and Coercive Practices

"Beware of false prophets, which come to you in sheep's clothing, but inwardly they are ravening wolves."—Matthew 7:15

"Natural selection builds child brains with a tendency to believe whatever their parents and tribal elders tell them . . . Such trusting obedience is valuable for survival ... But the flip side of trusting obedience is slavish gullibility."
—Richard Dawkins

Dishonest leaders, false prophets and fraudulent preachers who willfully deceive their followers are among the worst abusers of religious trust. Some of the greatest abuses have been perpetuated by Protestant Evangelists. Let me just mention a few names: Jim and Tammy Faye Bakker, Jimmy Swaggart, and Ted Haggard. I won't waste ink recounting their greedy and hypocritical activities, but in their prime, they were among this nation's most beloved televangelists. Before their fall from grace, Jim and Tammy Faye Bakker were taking in over a million dollars a week through their PTL Club. All these charlatans accumulated enormous tax-exempt wealth at the expense of their faithful but gullible followers.

David Koresh, head of the Branch Davidians of the Seventh-day Adventist Church; Jim Jones, a former Methodist who founded the Peoples Temple of the Disciples of Christ; and Marshall Applewhite, founder of the Heaven's Gate religion exercised the ultimate coercion when they led their followers to premature death. You may feel better by calling these latter religions "cults," but in reality there is no real difference between these "cults" and the presumably more respectable "churches." They are all peddling nonsense to pathetic, gullible fellow humans in order to gain control over them. In some respects, cults are the more impressive—they are able to win the minds of adults who in most cases had previously subscribed to a different belief. While cults must work to win converts, traditional religions have it much easier. They are able to indoctrinate children from infancy, and in many cases insulate them from any secular or other knowledge that might shake their faith. Many home-schooled American Christian children and madrasa-educated Muslim youngsters have been so thoroughly brainwashed that they will never understand that their manufactured worldviews are no more valid than the religious myths from which they are derived. Denying so many humans their intellectual potential may be one of religion's greatest crimes.

* * *

I would note that the Abrahamic religions maintain their hold over billions of followers through cruel threats and intimidation. All three faiths continue to employ various forms of coercion to insure obedience by their followers. Telling youngsters that religious disobedience will lead to eternity in the "Lake of Fire" serves to control them and keep them ignorant.[23] In all Abrahamic religions failure to conform/believe is punished by various other sanctions (ostracism, banishment, excommunication, etc). But of course the most extreme sanctions are seen in the Muslim nations of the Middle East and North Africa (especially Iran, Saudi Arabia, Somalia, Sudan, Libya, Nigeria and Afghanistan) where Sharia law prevails. Apostates and atheists may receive the death penalty. The majority of Muslims apparently approve of the death penalty for apostasy—a 2010 Pew Research survey revealed that 84 percent of Egyptian Muslims, 86 percent of Jordanian Muslims, and 76 percent of Pakistanis support the death penalty for leaving Islam. The notion that only a "tiny minority" of Muslims hold extremist views is nonsense—even in North America, Western Europe and the UK, significant numbers of second-generation Muslim immigrants still adhere to these extremist views.

Death sentences for apostasy are uncommon in today's world but certainly do occur—in May 2014 Meriam Ibrahim was sentenced to death by a sharia court in Sudan for refusing to recant her Christian faith. Fortunately, Ibrahim's sentence was later suspended because she was pregnant. Following that suspension, a colleague remarked that we were *weakening the shackles of religion ... one link at a time.* Not really!

The forms of coercion employed by world religions are too numerous to list here, but I'm sure the interested reader is aware of many. In the sections below we will be mentioning more of the mechanisms religions use to intimidate and control their followers.

IV. Embracing Cruel and Heartless Practices

> *"If a man commits adultery with another man's wife, both the man and the woman must be **put to death**." —Leviticus 20:10*

> *"A priest's daughter who loses her honor by committing fornication and thereby dishonors her father also, shall be **burned to death**." —Leviticus 21:9*

> *"Anyone who attacks his father or his mother must be **put to death**." —Exodus 21:15*

> *"Anyone who curses his father or mother must be **put to death**." —Exodus 21:17*

> *"If there is a man who lies with a male as those who lie with a woman, both of them have committed a detestable act; they shall surely be **put to death**." —I Corinthians 6:9*

Don't you find all that "putting to death" stuff a bit tiresome? If you have actually read your Torah, Bible or Koran, you must realize that those contemporary religionists who claim to interpret and obey their "sacred texts" literally are either lying or slightly mad! These are just a few of the capital offenses defined in Judeo-Christian scripture. In the West most people dismiss such heartless teachings as anachronistic and inappropriate. Unfortunately there are many throughout the world who still embrace such "laws" because they are certain that their scripture is divinely inerrant.

The public execution of Saudi Princess Misha'al in 1977 was documented in the 1980 film "Death of a Princess." Misha'al had dishonored her father by refusing to marry the old man chosen for her, and she ran off with her young lover. The couple received appropriate Islamic punishments for their wicked crimes—she was shot in the head, and he was beheaded after being forced to watch her execution. Actually the princess, presumably because of her royal status, was spared the more brutal punishment for adultery prescribed by Sharia law—public stoning. In the very religious regions of Iran, Afghanistan, Somalia, and Kurdish Iraq, women accused of adultery are sometimes subjected to public stoning. The woman is usually buried to her neck and the crowd throws stones at her head until she dies. (In contrast, men are usually buried only to their waist so they will endure much more injury and pain before they finally die—thus, stoning of women is considered more "humane" by apologists for the practice!) These are not isolated events or cultural aberrations—in 2013 Pew Research reported that 89 percent of Pakistani Muslims, 85 percent of Afghan Muslims, and 89 percent of Palestinian Muslims surveyed supported stoning for adultery. It is an extremely painful, horrific death that clearly constitutes torture. Stoning is also employed in "honor killings" in the Arab world. In 2007, Westerners videotaped the honor killing by stoning of a 17-year-old Kurdish girl who had an affair with a young man of a different religious group. In July 2010, a 43-year-old widowed Iranian mother of two was ordered stoned to death for adultery. She had confessed to the crime after receiving 99 lashes, but subsequently retracted her confession. On May 2014 in Lahore, Pakistan, 23-year-old Farzana Parveen, who was three months pregnant, was stoned and beaten with sticks by family members who were outraged that she had dishonored them by marrying against their wishes. She died on the pavement outside the local courthouse as police and local residents stood by and watched. Farzana's was not an isolated case—there were 869 such "honor killings" recorded in Pakistan in 2013. The vast majority were perpetrated by family members who had been "embarrassed" by the victims' actions.

Today, honor killings can be found closer to home. In Irving, TX on New Year's Day 2008, 18-year-old Amina and 17-year-old Sarah Said were shot dead by their father after he discovered that they had been dating non-Muslim boys. And the previous year in Mississauga, Ontario (a Toronto suburb), 16-year-old Aqsa Parvez was strangled by her father for her resistance to wearing a hijab, the traditional shoulder-length headscarf worn by devout Muslim women. In February 2011, Muzzammil Hassan was found guilty of second-degree murder for decapitating his estranged wife at the office of their television station in Orchard Park, NY, a Buffalo suburb. Some authorities have speculated on whether this was an honor killing and whether the couple's religion had a role in the crime. In January 2012, three members of an immigrant Afghan family in Kingston, Ontario were found guilty of first-degree murder in what was clearly an "honor killing." Mohammad Shafia, with the help of his wife and 21-year-old son, drowned his three daughters aged 13 to 19. Wiretaps played during the trial recorded the father calling his daughters *"filthy whores"* who *"betrayed"* his family by dating boys and wearing revealing clothing. At the same time, Shafia also drowned his first wife, who had failed to bear children but had still lived with them as a family member/servant.

On August 10, 2015, CBS News reported the tragic tale of a 20-year-old young woman in Dubai who drowned at the beach when her father prevented lifeguards from touching her (and thereby "dishonoring" her). The father became violent and physically pulled the would-be rescuers away while saying that he would prefer his daughter die rather than "be touched by a strange man." Dubai police did arrest him, but only after his daughter had died.

In some cases, Muslim clergy have come forward to argue that these deaths, whether "honor killings" or public executions, are the result of cultural, not religious practices. That is an absurd distinction—these acts are consistent with certain "Islamic cultures"—the cruel cultural milieu that prevails in areas governed by Sharia law is the product of a cruel religion! The obscene treatment of women throughout the Arab world is clearly prescribed by Islam's sacred laws. When Westerners criticize such brutality they are invariably accused of attacking Islam. When the film *Death of a Princess* was shown in the UK and US in 1980, Saudi Arabia threatened to expel our diplomats and cut off oil supplies. Not much has changed since.

* * *

The Taliban in Afghanistan and Pakistan still attempt to deny educations to girls and have been known to mutilate and/or murder teachers who oppose their will. And the Muslim-extremist Boko Haram in Nigeria continues to mutilate, murder and abduct citizens to prevent them from being exposed to education and "corrupt Western" (aka "non-Muslim") values. Since abducting more than 200 schoolgirls in April 2014, Boko Haram has continued to kidnap women and girls in its reign of terror.

Hateful teachings and practices are not restricted to the Abrahamic faiths. Hindus in parts of northern India frequently commit "honor killings" for reasons similar to Muslims. They, too, often prefer to consider it "cultural" rather than "religious," but I find the distinction unconvincing in a Hindu-dominated culture. Hindu law clearly prescribes the death penalty for women found guilty of infidelity. Equally heinous is the Hindu practice of court-ordered gang rape as punishment for cultural/religious disobedience. In January 2014, a 20-year-old woman was sentenced to "gang-rape" by a village court in eastern India—she was accused of having a relationship with a Muslim man from a nearby village. She was tied to a raised platform so the entire village could watch her being assaulted by the 13 "jurists" appointed to punish her. Her presumed lover was released after promising to pay a fine! Few villagers expressed any disapproval of the action—most said the woman needed to be punished. Rape is widespread but under-reported in India. Historically, upper-caste Hindu males have raped (and murdered) lower-caste women with impunity, but such rapes were rarely reported because the victims would be ostracized by their "disgraced" families and shunned in their villages. Only in the past few years have women been encouraged to report rapes—largely as a result of several widely reported gang-rapes of European tourists by groups of unemployed men. But the problem for native Indians continues—in May 2014, two teenage girls were raped and murdered by males of a higher ranking caste. Two police officers were said to be complicit. As I write this, India continues to struggle with a serious international image problem with almost monthly reports of gang rapes. Arguing whether such behavior is religiously or culturally motivated is absurd. If the people committing these atrocities justify their acts on religious grounds, as they almost always do, then the answer is clear. Indian political and social rights activist Arundhati Roy has recently addressed the caste system/rape crisis in India: *"The trouble is that once you see it, you can't unsee it. And once you've seen it, keeping quiet, saying nothing, becomes as political an act as speaking out. There's no innocence. Either way, you're accountable."* Few expect any immediate legal reform to suitably deal with India's problem.

On a dubiously positive note, I should point out that recent legal reforms in Turkey have increased the penalties to those involved in honor killings. Unfortunately there has been a concomitant increase in the number of "unexplained" suicides by young women. Apparently a young woman who has disgraced her family is locked in a room with a gun or a rope and is expected to redeem herself by taking her own life. The practice is referred to as "forced suicide."

Finally, while it is not my practice to defend Christianity, I am compelled to point out that Jesus brought a new paradigm to the West when he condemned public stoning in general, and for adultery in particular—*"Let he among you who is without sin, cast the first stone."* Unfortunately, the lesson has been lost on those fundamentalist Christians who still advocate the murder of abortion providers. Or that crazy Catholic priest in Poland who has urged that his country return to the medieval practice of burning homosexuals at the stake.

You don't have to look very hard to find other examples of cruel and heartless religious rhetoric and practice among fundamentalist Jews, Christians, Muslims and Hindus!

V. Promoting Harmful Beliefs and Strife

"The belief of a cruel God makes a cruel man." —Thomas Paine

"A good world does not need a fettering of free intelligence by words uttered long ago by ignorant men." —Bertrand Russell

And God said to them, "Be fruitful and multiply and fill the earth and subdue it and have dominion over the fish of the sea and over the birds of the heavens and over every living thing that moves on the earth." —Genesis 1:28

It is important for individual religions to discredit competing faiths to establish their superiority. To accomplish this, organized religions often require their followers to accept harmful teachings. Earlier we discussed the mutual intolerance of the major Abrahamic faiths. They teach malevolent myths about one another and end up generating uncontrolled hatred. Extreme intolerance also exists between the various sects within the individual faiths. A few familiar examples will suffice: Most of the suicide bombers in Iraq were Sunnis (often Al Qaeda members) attacking Shiites; the long conflict in Northern Ireland was Catholic against Protestant; the Bosnian war pitted Muslims, Catholics and Orthodox Christians against one another; the Israeli occupation of Palestine pits Jews against Muslims. You can think of more. Clerical teachings are often promulgated for the purpose of generating religious hatred and are directly responsible for untold deaths.

We are all too familiar with the pathetic acts of Muslim suicide bombers who willingly sacrifice their lives for Allah because they have been told that they will go directly to paradise as heroic martyrs and will be rewarded on their arrival with "seventy virgins".[24] Encouraging young men and women to commit murderous acts of self-sacrifice for a religious cause can only be described as evil. All the Abrahamic faiths consider suicide to be a serious sin—according to the very religious, your life belongs to God and it is his decision when it shall end. To them, taking your own life is no different than killing another. Of course, Muslim suicide bombers are acting to "defend the faith" and are thereby exempted from this restriction.[25]

Religious teachings are often harmful to the health of planet Earth. Many indigenous American and Eastern religions, especially the polytheistic faiths, teach a respect for nature, but such a concern is rarely evident in Abrahamic tradition. Many environmentalists believe that the Divine injunction in Genesis telling humankind to *"multiply and subdue,"* if not a direct cause, is at the least a major factor in the mindset that has led to global overpopulation and resource depletion.[26] An individual with an absolute belief in an apocalyptic "end time" and an afterlife is less inclined to concern himself with the health of the planet. If you think that is far-fetched I urge you to recall James Watt, President Reagan's first Secretary of the Interior. Watt, a born-again Christian, believed that the *"Second Coming was at hand"* and felt we would be wise to exploit all our natural resources while we were able. He is reputed to have said, *"After the last tree is felled, Christ will come back."* The authenticity of that exact quote has been challenged—Watt later denied having said it, but it is certainly consistent with his and Reagan's environmental policies. In testimony before the House Interior Committee on February 1981, Watt described his responsibilities thus: *"That is the delicate balance the Secretary of the Interior must have: to be steward for the natural resources for this generation as well as future generations. I do not know how many future generations we can count on before the Lord returns."* In May of that year, the *Washington Post* recorded him saying: *"My responsibility is to follow the Scriptures which call upon us to occupy the land until Jesus returns."* He summarized his goals when he added: *"We will mine more, drill more, cut more timber."*

As noted earlier, Americans remain remarkably misinformed on environmental issues. Surveys by Ipsos MORI (2014), one of the UK's leading social research organizations, reveal that Americans are pathetically ignorant regarding global climate change. Only 54 percent of us believe that global warming is the result of human activity. In contrast, 93 percent of Chinese, 80 percent of Indians, 79 percent of Brazilians and even 67 percent of gas and oil-dependent Russians acknowledge that global climate change is anthropogenic. We are a nation of "deniers," and the vast majority of those deniers are fundamentalist

Christians. They have been taught that humankind is not capable of destroying the planet—that is a privilege reserved for their god. Many of the Religious Right actually believe that the environmental movement is a threat to Christianity.

* * *

Of course, the greatest environmental harm comes from the unacknowledged global overpopulation that results from the three Abrahamic faiths' opposition to birth control, consistent with the injunction *"to multiply."* This common belief that birth control is sinful is derived from God's execution of Onan for failing to properly inseminate his widowed sister-in-law.[27] The Catholic Church obsesses about sex and birth control more than any other contemporary religion. In fact, agents of the church have been known to deliberately deceive their followers in order to discourage use of contraceptive technology. One glaring example comes to mind—Catholics in the Philippines have been told by representatives of their church that condoms don't protect against the HIV virus. Why would a Catholic Cardinal spread such an obvious falsehood? Apparently because he would prefer that his subjects contract AIDS rather than commit the mortal sin of contraceptive use—they can still get into Heaven if they contract AIDS but they will surely spend eternity in Hell if they commit a mortal sin and fail to have it properly absolved and forgiven. We will have more to say about the perverse sexuality promoted by Western religions below.

VI. Inhibiting Spirituality and Moral Progress

"It seems to me that the path to genuine religiosity does not lie through the fear of life, and the fear of death, and blind faith, but through striving after rational knowledge."
—*Albert Einstein*

"We keep on being told that religion, whatever its imperfections, at least instills morality. On every side, there is conclusive evidence that the contrary is the case and that faith causes people to be more mean, more selfish, and perhaps above all, more stupid." —*Christopher Hitchens*

There is a lot a talk about "spirituality" today, mostly by people who hope to use it for some sort of personal gain. There are "spirituality workshops" to help you improve your mind, lose weight, improve your sex life, earn extra income, and all sorts of other mundane purposes. Among some of the more erudite, spirituality has come to replace religiosity and is often associated with the wisdom of Eastern religions. We commonly hear seemingly sophisticated people say, "I'm not religious but I'm spiritual," without making clear what they mean by either term. There are no consistent definitions, so allow me to offer mine. I consider a person religious if they have "faith" in the existence of a personal god or god-like figure and they adhere to some form of worship of that deity. Religious people may or may not be spiritual. I will be using the term spirituality in referring to a sense of transcendent connectedness with a larger reality that the mature human mind can accomplish. Spirituality does not require "faith" or an irrational belief in a supreme being. The spiritual individual often gains an enhanced appreciation ("understanding") of his or her own essence. One needn't be intellectually sophisticated to be spiritual—many primitive peoples have a profound connectedness to their environment—think of the shepherd 4,000 years ago guarding his flocks while studying and discovering patterns in the constellations of the night sky—or the pre-Columbian Native American kneeling over the deer he has just killed and conversing with its spirit. In my mind, these two "primitives" enjoyed a spirituality that was richer than what most of us "moderns" know. With no formal education they examined their world and sought deeper understanding about their existence. In contrast, the person who accepts a worldview devised for them by an organized religion and mindlessly memorizes passages of scripture attains a very shallow spirituality at best. Attainment of authentic spirituality is often thwarted by formal religion.

Spirituality is not necessarily a connection with "the divine" as your religion may want you to believe—it can be, for example, a connection with one's culture. It usually begets a deeper understanding of self that is gained through an introspection that integrates our life experiences with the world of knowledge. In today's complex and changing world it normally requires lifelong reflection—a practice often discouraged by formal religion.

The reader should understand that there is no conflict between science and spirituality. Carl Sagan stated it better than I can: *"Science is not only compatible with spirituality; it is a profound source of spirituality. When we recognize our place in an immensity of light-years and in the passage of the ages, we grasp the intricacy, beauty and subtlety of life, then that soaring feeling, the sense of elation and humility combined, is surely spiritual. . . . . The notion that science and spirituality are somehow mutually exclusive does a disservice to both."*

Meditation, contemplation, study and even prayer can be vehicles to spirituality. Unfortunately, when religions convince youngsters that the "religious awe" they experience is spirituality, they will end up substituting a "brain gimmick" for real spiritual development. Their "feel-good" religious awe is nothing more than the misappropriated physiological reactions that characterize interpersonal love—a conditioned release of neurotransmitters, especially dopamine, in the brain—much like the response to certain recreational drugs. And like recreational drugs, "religious awe" can ofttimes be addictive and harmful. When young people accept "religious awe" as the essence of their spiritual nature it will likely impede their attainment of higher levels of understanding and morality.

Most Western scholars subscribe to the notion of "moral progress"—that is, they believe that over time our species has undergone progressive social evolution that has led to increasingly moral cultures—we have become more humane and charitable through the course of recorded history. Most of us would agree that humankind today is, on average, less cruel and more generous and fair than were our ancestors of several centuries ago—we use the term "moral progress" to refer to the perceived changes in human ethical behavior over time. In his recent book *The Moral Arc* (Henry Holt & Co., 2015) Michael Shermer argues that it has been the widespread adoption of science and reason that has led us on a path to greater justice and freedom. Like many of us, Shermer sees institutionalized religion as the major impediment to both individual and collective "moral progress." I have neither the time nor the inclination to review all the evidence in support of his arguments—the interested reader can either consult the published literature (e.g., Shermer's thorough review) or just use a little common sense. You only need to examine post-Neolithic human cultures, especially those dominated by the Abrahamic religions, and you will find widespread acceptance of exceedingly cruel practices that today we would consider clearly immoral. Most reasonable people would see our abandoning those cruel practices as part of our moral progress. Not everyone would agree—Fundamentalist missionaries see the successful conversion of heathen "natives" to Christianity by means of religious indoctrination as "moral progress"—I see it as the exact opposite!

The history of our country may be more instructive—consider Manifest Destiny and the fate of our Native American predecessors. Or our widespread enslavement of African Americans—or the post-Emancipation "American apartheid" wherein many whites continued to segregate and economically and politically oppress the nonwhite population. That so many of us have abandoned those practices is evidence of our moral progress.

I think that globally, on average, we can show a gradual improvement of the human moral condition. But as Shermer asserts, it has been widespread adoption of science and reason, and not religious indoctrination, that has led to greater justice and freedom—and consequent improved morality. I grant little credit to contemporary religions for that progress.

VII. Encouraging Irrational Thinking

"I do not feel obliged to believe that the same God who has endowed us with sense, reason, and intellect has intended us to forgo their use." —*Galileo Galilei*

"Say what you will about the sweet miracle of unquestioning faith, I consider a capacity for it terrifying and absolutely vile." —*Kurt Vonnegut*

"Once God is accepted as the first cause of everything which happens in the mortal world, nothing is left to chance ... logic can be happily tossed out the window." —*Stephen King*

"Faith means not wanting to know what is true." —*Friedrich Nietzsche*

"In classical Greek philosophy, faith was the lowest grade of cognition ... the state of mind of the uneducated." —*E.R. Dodds*

"Faith may be defined briefly as an illogical belief in the occurrence of the improbable. A man full of faith is simply one who has lost (or never had) the capacity for clear and realistic thought." —*H.L. Mencken*

As I noted earlier, Western religions are in a never-ending war with science. The basis of the conflict is obvious—science relies on reason, whereas our religions require faith. More precisely, science seeks to offer tentative explanations to natural phenomena with no expectation of finding "absolute truths." Scientists and scientifically literate people understand that their knowledge base is self-correcting and always evolving. Religions, on the other hand, claim access to the divine and eternal "absolute truths" found in their inerrant sacred texts, and they require their followers to accept those "truths" through unquestioning faith alone. Western religions have traditionally taught that when a conflict arises between science and Scripture, the faithful must follow Scripture until their church leaders allow otherwise. Contrast this with Buddhism, wherein the follower is urged to question. Five hundred years before the birth of Christ, the Buddha said:

> *Do not believe in anything simply because you have heard it. Do not believe in anything simply because it is spoken and rumored by many. Do not believe in anything simply because it is found written in your religious books. Do not believe in anything merely on the authority of your teachers and elders. Do not believe in traditions because they have been handed down for many generations. But after observation and analysis, when you find that anything agrees with reason and is conducive to the good and benefit of one and all, then accept it and live up to it.*

Can you imagine an evangelical Christian minister, a Catholic priest, a Muslim imam or a conservative rabbi ever saying anything like that? I surely can't. Consistent with religious suppression of inquiry and knowledge is their disregard for intelligence. As noted elsewhere, Bertrand Russell pointed out that nowhere in Abrahamic scripture is there anything resembling praise of intelligence!

The fundamental decision remains, as always: how shall we choose to understand the universe—through science or religion? Excuse my bluntness, but the god of scripture is really quite ignorant regarding science. In fact any youngster knows as much about the universe as that god does—and give that child a simple microscope or telescope, and he or she will instantly know more than God. With a microscope the child will see a host of tiny life forms that God never told us about and apparently never knew existed. With a telescope the child will see a million-fold more and different "stars" than the god of scripture told us about. If you insist that He knew about them but didn't bother to tell us, I would point out that He did go to great lengths to tell us about a lot of creatures and heavenly events that did not exist! The Behemoth and the Leviathan in the Book of Job (Job 40:15 and 41:1) are bizarre mythological beasts that quite certainly never existed (despite the ridiculous suggestion of some modern evangelicals that they are the Biblical records of dinosaurs).

If God was all-knowing, why didn't he share some facts about the universe that we didn't already know? If he really wanted to convince us of his omniscience he could have given his correspondents something really significant—for example, he could have said, *"Energy is equal to mass times the speed of light squared,"* or something easier, like *"equal volumes of gases at the same temperature and pressure contain the same number of molecules regardless of their chemical nature and physical properties,"* (Then when we discovered it in the early 19th century we could have called it "God's Law" instead of "Avogadro's Law"—which would have been much easier for beginning chemistry students to spell.) Of course I'm being facetious, but think about it— wouldn't an all-knowing god have a little bit better information? I can find nothing in Scripture that convinces me that god is anything other than the invention of ignorant and frightened ancient people.

Unfortunately, as human understanding of the universe was expanded though reason and science, religions felt it necessary to discredit science. Our Western religions have intentionally indoctrinated their followers to disrespect science. Read the words of Martin Luther "explaining" Genesis: *"But, if you cannot understand how this could have been done in six days, then grant the Holy Spirit the honor of being more learned than you are."* Luther is telling you that if it doesn't make sense, just shut up and believe —ignore your sense of reason! It is much the same today when scientifically illiterate fundamentalists attempt to repudiate what they call "Darwinism." Our extensive knowledge of the physical, chemical and biological events underlying organic evolution overwhelms their absurd Creation myths from Genesis, and they can only hope to retain their followers by disparaging modern science. Personally, I consider it immoral to force your followers to embrace defective older models of the universe when there are contemporary models with far greater predictive value and accuracy. Unfortunately, the intellectual divide between science and religion promises to deepen with the accelerated progress of modern science.

VIII. Promoting Anachronistic and Perverse Morality

"The most curious social convention of the great age in which we live is the one to the effect that religious opinions should be respected." —*H.L. Mencken*

"If a man would follow, today, the teachings of the Old Testament, he would be a criminal. If he would strictly follow the teachings of the New, he would be insane."
—*Robert G. Ingersoll*

"Christianity: If it feels good, stop it!" —*Anonymous*

Rather than review all the nonsensical morality taught by the Abrahamic faiths I will address just one outstanding absurdity—their perverted obsession with sexuality. Islam and Christianity have distorted human sexuality into something quite unnatural. Fundamentalist Christianity equates sex with sinfulness, while contemporary Islam uses sex to manipulate society. All three Abrahamic faiths use perverted sexual morality to subjugate women.

The roots of this sexual obsession are not entirely clear, but there is little doubt that Western religions have inflicted guilt and psychological repression upon those who fail to adhere to their narrowly prescribed sexual practices. In contrast, Eastern religions tend to be more accepting of sexual expression and treat lovemaking as a natural act. Virtually every pre-Abrahamic polytheistic culture worshipped gods and/or goddesses representing sexuality and fertility, and usually treated sex as a normal human activity. We suspect that the limits imposed on sexual freedom by the monotheistic Abrahamic faiths in part reflect their attempts to distance themselves from the older polytheistic religions. As noted earlier, historical attempts to eliminate references to Asherah, the fertility goddess who was also Yahweh's wife, served to reduce the emphasis on sex at the same time reinforcing monotheism.

The monotheistic faiths denounced many earlier practices as sinful and licentious. Christians continue to look with disdain on the liberal sexuality of many Hindu sects, and are horrified by the public displays of phallus worship by some Buddhist sects. I expect that many Buddhists and Hindus view Western sexual mores with equal disdain. Space does not permit me to detail the extensive and often contradictory teachings about sex found in Muslim scripture. I will focus mainly on Judeo-Christian traditions.

We all know that the Old Testament contains numerous rules regarding sex, especially adultery and sodomy, and that these prohibitions become further expanded in the New Testament. In fact, when the Catholic Church saw fit to "alter" the original Ten Commandments they were able to place even more emphasis on sex.[28] Curiously, Jesus didn't have a lot to say about sex—but some of his followers, especially Paul and Matthew, surely did. Paul, of course, never knew Jesus and probably knew little about his actual teachings, so he was free to define Christianity as he wished. Unfortunately, his sexually repressive, misogynistic, and homophobic writings established early Christian policy and still influence it today. Retired Episcopal Bishop John Shelby Spong suggests that St. Paul's conservatism stemmed from the fact that he (Paul) was *"a self-loathing and repressed gay male."* Ironically, many contemporary Christians, suffering from the homophobia that Paul's writings have instilled in them, refuse to even discuss the hypothesis that their "first great teacher" might have been gay. (Bishop Spong makes a practice of offending conservative Christians in a way not unlike how Paul's writings have insulted women and gays over the centuries.)

The development of Christian attitudes toward sex through history is both fascinating and bizarre. Reay Tannahill, in her scholarly study *Sex in History* (Stein and Day, 1980) reminds us that *"early Christian leaders made sex and sin synonymous."* Much can be attributed to the writings of Paul, who lectured on everything from circumcision, virginity, celibacy, fidelity, marriage, incest, homosexuality and sodomy. In his writings Paul sounds very much like the sexually repressed gay male that Bishop Spong described. Matthew (19:12) reinforced Paul's ranting but went a step further in his special treatment of celibacy and castration—*"For there are some eunuchs, which were so born from mother's womb: and there are some eunuchs, which were made eunuchs of men: and there be eunuchs, which have made themselves eunuchs for the kingdom of heaven's sake.*" Origen of Alexandria (AD 185–254), considered one of Christianity's greatest theologians, was inspired by Matthew and others to castrate himself *"for the kingdom of heaven's sake."* Self-castration actually became fairly common among early Christian clerics until outlawed by the Council of Nicaea in AD 325. Tertullian (ca. AD 160–220), the prolific and extremely influential early Christian author, described both Jesus and Paul as *"spadone,"* a term which is variously translated as "eunuch" or "virgin." Elsewhere in his writings, Tertullian specifically refers to Paul as being *"castrated."* At the very least, *"spado"* can be considered a metaphor for celibacy. Of course, Tertullian saw these as noble traits and was being complimentary. (Tertullian is remembered as the first Latin writer to use the term *Trinity*—his description of heaven as a place where the virtuous would enjoy watching the eternal tortures of sinners is less well celebrated.)

Thus, the early Christian Church came to view celibacy as virtuous. In AD 386, when Pope Siricius decreed that Church elders should not have sex with their wives, he was largely ignored. But as the Church became increasingly controlled by celibate men, its attitudes about sex worsened. Reay Tannahill advises us: *"It was Augustine who epitomized a general feeling among the church fathers that the act of intercourse was fundamentally disgusting ... Arnobius called it filthy and degrading, Methodius unseemly, Jerome unclean, Tertullian shameful, Ambrose a defilement."* Of course, women became the object of their holy disdain—Hugues de Saint-Cher, a medieval cardinal, wrote: *"Woman pollutes the body, drains the resources, kills the soul, uproots the strength, blinds the eye, and embitters the voice."* Misogynistic priests came to preside over barbaric witch hunts throughout 15th to 17th century Europe—women accused of being morally corrupt were judged to be witches and often tortured and burned. Any women outside the close religious control of the local church were at risk—and the Church murdered many thousands. It has been suggested that the Church "invented" witches and demons in order to explain evil in a world controlled by an omnipotent and loving god and thereby rationalize the Theodicy Problem (see note **8**).

In recent centuries, regional and national governments, influenced by prevailing religious beliefs, have passed ordinances regulating personal sexual behavior. In the United States, laws relating to adultery and homosexuality became common. Many remain on the books—as recently as 2010, a woman in the Buffalo area was arrested and convicted of adultery, ostensibly for having sex with a man other than her husband. (In fact, she was actually having drunken sex on a picnic table with a man half her age in a busy public park at midday—that strikes me as something more than adultery!) So-called "sodomy laws" prohibiting "unnatural" (e.g., anal and oral) sexual acts were still on the books in 13 states in 2003. In AL, FL, ID, LA, MS, NC, and SC, "sodomy" was illegal for everyone—straight or gay. In KS, MO, OK, and TX, "sodomy" was criminal only when performed by members of the same sex, so their laws were really about homosexuality. In June 2003 the US Supreme Court determined that all sodomy laws were unconstitutional and that adults were free to engage in consensual sex in private without government oppression. Most of the sodomy laws in the Southern states were overturned at that time. (Despite the high court ruling, Republicans in the Michigan Senate in early February 2016 passed a bill again outlawing sodomy—*"the abominable and detestable crime against nature with mankind or with any animal."* This was the same week that the rest of the nation and world were watching that state's failure to adequately deal with the toxic lead levels in the municipal water supply in Flint.)

The Christian Church has a long history of opposing birth control, and by the Middle Ages taught that sex for any reason other than procreation was immoral. Taking pleasure in sex, especially by women, was especially sinful. It was apparent to the clergy that birth control could lead to unregulated sexual freedom and pleasure, so birth control by any means, including abortion, was prohibited. As noted earlier, the Church's rationale for most of these prohibitions was based on the story of Onan, whom God killed for failing to inseminate his brother's widow.

The "Sin of Onan" was the name given to *coitus interruptus* (premature withdrawal), the only form of contraception available to the poor throughout most of history. Wealthier people could always purchase widely available contraceptive and/or abortifacient botanicals as well as various simple barrier devices. The Roman Catholic Church, however, rigorously banned all means of contraception. In 1930 Pope Pius XI explicitly said that married people should have sex with the full expectation that children could result each time, and to circumvent that expectation was a grave sin. About that time, reproductive biologists showed that ovulation occurred mid-cycle and established that pregnancy could only occur if the woman was inseminated near that point. The Church reluctantly acknowledged that avoidance of sex during the days surrounding ovulation would decrease the likelihood of conception and might be an acceptable form of birth control. But the reasoning they used to rationalize the use of the "rhythm method" is amusing. The Church had long insisted that celibacy was a good thing because the clergy was expected to practice it. Therefore, a little sexual restraint mid-cycle by the couple might also be good. It was fine with the Church as long as the couple was merely practicing periodic celibacy and not conspiring against God to prevent birth. (The "Ceiling Cat God" is watching so don't try to fool him!) The acceptance of the rhythm method by the Church caused H.L. Mencken to quip: *"It is now quite lawful for a Catholic woman to avoid pregnancy by a resort to mathematics, though she is still forbidden to resort to physics or chemistry."*

In 1968 Pope Paul VI assembled an advisory panel of 72 prominent Catholics including seven cardinals, nine bishops, 16 theologians, 13 physicians and 27 lay people (the panel actually included several women!) and asked them to help sort out the "problem of contraception." Following due deliberation, 65 of the 72 (~90 percent) concluded that contraception *was not* morally evil. In his Divinely inspired wisdom, Paul VI then issued his encyclical *Humanae Vitae*, proclaiming that contraception *was* morally evil. So much for papal consultation. Hundreds of US priests signed a letter to the *NY Times* protesting the decision. Four seminarians from Buffalo signed the protest, and three were immediately dismissed. The fourth was a moral philosopher who explained he was retained because *"they need me here."* Obviously.

Early religious pressures in the United States caused many states to pass laws prohibiting birth control. Connecticut law specifically prohibited the use and sale of *"any drug, medicinal article or instrument for the purpose of preventing conception."* It wasn't until 1972 that the US Supreme Court finally overturned state birth control laws, ruling that they violated the *"right to marital privacy."* For a time Massachusetts' laws remained because they only denied contraception to unmarried couples. But later in 1972 the courts found that law unconstitutional as well and established that the right of privacy applied to all citizens, married or not.

I find that episode in our history frightening because our churches, acting through the state, sought to control the most private and intimate aspects of our lives. But I find the same actions today with respect to abortion choice equally invasive. All of the Abrahamic faiths have historically sought to control what women can do with their own bodies, and some still teach that a woman's body (and especially her genitalia) belong not to her, but to her husband through the marriage contract. Such suppression of women's rights remains a serious problem in many contemporary Jewish, Muslim, Hindu and Christian cultures. I don't need to itemize the abuse that women suffer in many Arab countries. Some progressive Muslims will argue that their god treats men and women equally, but much of their scripture says otherwise. Their prophet has said, *"Among the inmates of Heaven women will be the minority,"* and *"I have seen that the majority of the dwellers of Hell-Fire were women ... they are ungrateful to their husbands and they are deficient in intelligence."* It would appear there is little place in Muslim Paradise for the fairer sex, except for those sweet virgins who will serve to pleasure the male martyrs of the faith.

IX. Wasting Material Resources and Human Potential

"I finally decided that I'm a creature of emotion as well as of reason. Emotionally, I am an atheist. I don't have the evidence to prove that God doesn't exist, but I so strongly suspect he doesn't that I don't want to waste my time." —Isaac Asimov

"Just in terms of allocation of time resources, religion is not very efficient. There's a lot more I could be doing on a Sunday morning." —Bill Gates

"When his life was ruined, his family killed, his farm destroyed, Job knelt down on the ground and yelled up to the heavens, "Why god? Why me?" And the thundering voice of God answered, "There's just something about you that pisses me off." —Stephen King, "Storm of the Century"

If you are a believer, just for a moment try to imagine that there is no Supreme Being.[29] Now let's examine the things done in "His Name" that waste precious lives and resources. How many human hours are wasted in pursuit of His approval? Worldwide every year, hundreds of thousands of men and women leave mainstream society to spend their lives in silent and seemingly endless prayer, "study," and contemplation. The cloistered Christian nuns and monks are familiar examples but there are similar Jewish, Muslim, and Hindu ascetics throughout the world. In Israel, for example, there are about 800,000 ultra-Orthodox Haredim, many of whom subsist principally on welfare. Haredim men refuse to participate in the military and may avoid entering the workforce so that they may spend their lives in full-time study of the Torah. They currently comprise nearly 10 percent of Israel's population but their number is rapidly increasing because their fundamentalist beliefs encourage large families.

(For the record, prayer doesn't work—numerous carefully controlled scientific studies document that fact. Prayer simply allows the practitioner to feel like they are doing something useful, when in fact they are only talking to themselves. It lets them feel good about doing nothing. In my opinion, public prayer as widely practiced by fundamentalist Jews and Hindus is both hypocritical and obscene—Christians might want to read Matthew's (6:5–6) thoughts on the matter.)

If there is no Supreme Being, these various religious extremists really contribute nothing to the planet, and in most respects consume far more than they produce. Most are supported by charity, although some in this country partially subsidize their existence by exploiting their tax-exempt status to market various products that may or may not be results of their labor. Consider further those wretches who feel compelled to placate their Supreme Being through self-deprivation and/or self-torture. Add to these the numerous isolated communities of religious fundamentalists who shun all contact with the modern world. Finally, add the many thousands of missionaries who spend some or all of their lives attempting to convert "heathens" to their particular faith. The cumulative person-hours spent on "serving the Lord" are incalculable. If we continue to assume there is no Supreme Being, then most of those lives represent an enormous loss of human potential. You may argue that the modest resource requirements of these "holy servants" are of little consequence, but if they collectively contribute nothing to the planet other than their prayers, we would still be better off without them. I will acknowledge that some great works of art have been produced by inspired ascetics, but most religious art has been produced in response to compensation rather than inspiration—traditional artists were commissioned by wealthy patrons who were hoping to buy their way into heaven. A similar argument applies to the sciences—little of value has been produced by religious ascetics. Before the middle of the 19th century many scientists were religious, but religion was not normally the source of their inspiration. Out of respect for my mother I must cite one of her favorite cases "proving" the compatibility of science and religion—Gregor Mendel performed his novel genetic studies on garden peas while he was an Augustinian monk—one of the few examples of important scientific research not thwarted by the Church. Unfortunately, Mendel never made his findings widely known to the scientific community, and his ideas languished for half a century until "re-discovered" by mainstream researchers. I'm not sure he actually deserves any credit—*"if a tree falls in the forest and nobody hears it. ..."* Mendel, by the way, was no ascetic and by the time he was 47 had abandoned scientific research and teaching to assume the largely administrative duties of abbot of the monastery.

When it comes to acquiring and wasting material treasure, the organized churches have few peers. The Roman Catholic Church requires great wealth to support its lavish displays. (As noted elsewhere, the Catholic Church was obliged to "re-write" the original Second Commandment in order to justify their idolatrous art and architecture: *"You shall not make for yourself an idol in the form of anything in heaven above or on the earth beneath or in the waters below"* [see note **28**].) The Catholic Church is reputed to be the world's single largest "corporation," enjoying an annual income from donations that approaches all the tax revenue of the US government. In addition, the Church has enormous real estate holdings—the Archdiocese of Boston alone owns over a billion dollars' worth of property. There are 121 other dioceses in just this country, so you do the math. Less well known are their worldwide investment holdings. In the US, the Church is one of the largest players on Wall Street. The Vatican also owns many billions in gold bullion stored in American, English and Swiss banks. The Vatican's unique status as both a sovereign nation and a tax-exempt entity enables it to conceal and preserve its vast wealth. It is impossible to even estimate the net worth of the Church, but some experts suggest it is the wealthiest organization in the world. Unfortunately, the Church is traditionally niggardly in its support of the world's poorest citizens and mostly uses its assets to maintain the status quo.

Much of its wealth is dissipated in lavish displays designed to impress and retain the faithful. I remember as a young man traveling through the small towns of central Québec and being awed by the extravagant architecture of the cathedrals. Most men in those towns earned a poverty-level income cutting pulpwood, yet they and their families were happy to support those outlandish edifices of prayer. Few children completed high school, and only a rare individual enjoyed any legitimate higher education. Without realizing it, they were literally enslaved by their faith. Great houses of worship around the world consume vast physical and human resources for their construction and maintenance. If there is no Supreme Being, that would seem a terrible waste.

Religions often distract us from the things that are really important and cause many of our fellows to invest their lives and resources in meaningless, even negative, activities. Vast numbers of people throughout history have been convinced that they must go to ridiculous lengths to appease their gods. We have all seen images of Hindu and Muslim religious fanatics in acts of self-mutilation—the pain apparently provides a penance for the imagined sins of mankind. They may spend hours in meaningless rituals—Jews "praying" at the Wailing Wall, Hindus, Muslims and Catholic monks self-flagellating, or devout Muslims undertaking lengthy pilgrimages to visit holy shrines. Any reasonable and decent god would want his people to make better use of their time and treasure.

X. Perpetuating Injustice and Oppression

"The ultimate tragedy is not the oppression and cruelty by the bad people but the silence over that by the good people."
—*Martin Luther King, Jr.*

"I have been sure of many things which were not true."
—*Oliver Wendell Holmes*

"All right, let's admit it; we Jews killed Christ – but it was only for three days." —*Lenny Bruce*

Perhaps you read the following Associated Press release by Nicole Winfield (*Washington Post*, 2 Mar, 2011) in your newspaper:

> *VATICAN CITY -- Pope Benedict XVI has made a sweeping exoneration of the Jewish people for the death of Jesus Christ, tackling one of the most controversial issues in Christianity in a new book. In "Jesus of Nazareth-Part II" excerpts released Wednesday, Benedict explains biblically and theologically why there is no basis in Scripture for the argument that the Jewish people as a whole were responsible for Jesus' death. Interpretations to the contrary have been used for centuries to justify the persecution of Jews.*

Wow! Pope Benedict absolved the Jewish people *"as a whole"* for the death of Jesus Christ. As my former Hebron Academy students would have sarcastically commented: *"That's pretty white of him!"* Benedict's exoneration fails to undo the immeasurable harm caused by two millennia of hateful Catholic teaching. Slanderous speech, whether by past popes or prep school boys, has consequences. It is true that for the past five decades the Church has taught that the Jews were not collectively responsible for the death of Jesus, but that belated message enjoyed limited circulation and could not reverse the centuries-old stigma labeling Jews as "Christ-killers." We know that Adolph Hitler believed he was acting in accordance with Christian doctrine and that the Holocaust was the direct expression of his belief. Of course Hitler didn't act alone— millions of other Christian Europeans, not just Germans, shared his anti-Semitic views and were, by virtue of their beliefs, complicit in the deaths of millions of Jews and others. Hitler may have been a madman but he was not an idiot—he would not have initiated a massive genocide if he thought it would be abhorrent to and opposed by a great majority of his German people.

Much of Scripture was contrived to discriminate, and as such tends to perpetuate injustices based on hatred. Elsewhere I've cited considerable examples from Biblical and Koranic text that reveal such discrimination. As a change, let me consider an example from another sacred source—the Book of Mormon (BOM). That text contains what can only be called outrageous racist doctrine. In 2 Nephi 5:21, BOM we read: *"And God had caused the cursing to come upon them, yea, even a sore cursing, because of their iniquity. For behold, they had hardened their hearts against him, that they had become like unto a flint; wherefore, as they were white, and exceedingly fair and delightsome, that they might not be enticing unto my people, the Lord God did cause a skin of blackness to come upon them. And thus saith the Lord God; I will cause that they shall be loathsome unto thy people, save they shall repent of their iniquities."* That passage describes how God, angry with the some of the descendants of the "lost tribe," cursed them and turned their formerly *"delightsome"* white skin to *"loathsome"* black. The remaining good white people were the Nephites, and the unbelieving others were the Lamanites. The blackness was placed on the Lamanites so that the Nephites could easily identify them and that *"they might not mix"* with them. The Lord specifically forbids miscegenation between the Lamanites and the Nephites (2 Nephi 5:23). The Lamanites were said to be *"a filthy people, full of idleness and all manner of abominations"* (BOM, 1 Nephi 12:22-23). Eventually the bad (dark) Lamanites destroyed most of the good (white) Nephites. It is generally agreed that Native Americans represent the descendants of the Lamanites. Elsewhere we learn that God inspired the European settlers, who were described as *"white, and exceedingly fair and beautiful"* to slaughter the dark-skinned Native Americans (1 Nephi 13:15).

Despite 19th- and 20th-century efforts to undo it, the pro-white discriminatory perspective is still in evidence. The

Mormon attitude toward African Americans remains ambiguous and problematic. The early Church actually attempted to adopt an anti-slavery position. In 1833 Joseph Smith dictated the Covenant (101:79) opposing slavery: *"It is not right that any man should be in bondage to another."* That same year, W.W. Phelps, Assistant President of the Church, wrote a letter published in the Church's newsletter inviting free blacks to come to Missouri and become Mormons. Unfortunately, the good Christian people of Jackson County, MO didn't much like the idea of "Yankee abolitionists" inviting all those black people to join them. Various disagreements between the "Old Settlers" and Mormons led to the "Missouri Mormon War" and the eventual expulsion of the Mormons from the state. By 1835, however, the Mormon Church had modified its position on slavery saying that if the United States government allowed it, they would not *"interfere with bond-servants, neither preach the gospel to, nor baptize them contrary to the will and wish of their masters, nor meddle with or influence them in the least to cause them to be dissatisfied with their situations in this life, thereby jeopardizing the lives of men"* (Covenant 134:12). In support of this position, Phelps reexamined biblical mythology and adopted the story that Egyptus, the reputedly black wife of Ham (son of Noah), was descended from Cain and transmitted *"the curse of Cain"* to their post-diluvian descendants. In addition, it was claimed that Ham himself was cursed with black skin for marrying a black woman. To those early Mormons, the *"Mark of Cain"* was black skin. While the Mormon Church claims that they have never overtly discriminated against blacks, they did not allow them into the priesthood until 1978—if a man fails to join the "Melchizedek priesthood" he will never attain the highest level of glory in the afterlife. It is really quite amazing that so many Pacific Islanders and blacks, especially Africans, subscribe to the Mormon faith. I must assume they don't actually read the Book of Mormon!

The Mormons may appear conflicted in their stance on race, but they have made their position on homosexuality very clear—traditionally they have ranked it right after murder as a most heinous crime. Or at least they did until it was no longer politically expedient. In 2008 they joined with the Catholics and Evangelical Christians to support passage of California Proposition 8 to disallow same sex marriage. But by 2013 the Mormons realized they were losing membership and seriously damaging their public image, so they began to soften their position on homosexuality. However in the last year or so, they seem to be reverting to their former hard-line position on same sex marriage! It's interesting how quickly a church can reverse its practices when their public image is threatened.

Homosexuality is, of course, another area where organized religions perpetuate injustices that lead to persecution of a minority group of people. Christianity has traditionally condemned homosexuality despite the fact Jesus never had anything to say on the subject. Why would he, if *"some of his best friends"* (e.g. the apostle John) were probably gay? The few examples purporting to show Jesus' views on male homosexuality (e.g. Matthew 19:4) are dubious interpretations at best. There is little doubt that Paul did condemn homosexuality, but as we have pointed out earlier, Paul had no direct knowledge about what Jesus actually said or thought and was probably on his own guilt trip imposed by homophobic Hebrew culture at the time. Irrespective of what Jesus may have thought, mainstream Christianity, following the perverse teachings of Paul and Matthew, came to preach that homosexuality was evil. Today, the most progressive and enlightened Christian churches claim to treat homosexuals equally to heterosexuals. Despite having knowingly ordained many gay men, the Catholic Church does not at this time condone gay marriage. They reason that the vows of celibacy make priests' sexual orientation inconsequential, whereas married gay couples will actually perform an act that they still consider a grave sin.

Traditional Judaism was very explicit in its condemnation of male homosexuality (Leviticus 18:22; 20:13) and treated it as a capital crime. Today, on the other hand, the more liberal branches of Judaism permit the ordination of gay and lesbian rabbis and allow rabbis to perform same-sex commitment ceremonies. This, however, has been a very contentious issue for various other Jewish groups—contemporary Jews, like Christians, seem to span the full spectrum of opinions on the issue.

Most Muslims remain adamant in the view that humans are naturally heterosexual and that homosexuality is sinful and unnatural. In Islamic law both male and female homosexuals are punished by severe beatings—repeat offenders are stoned to death. (Yet the Koran describes the martyr's reward of *"boys of perpetual freshness."* Pederasty is condoned, but adult homosexuality is a capital offense. Go figure!)

One of the problems in rationally discussing homosexuality is establishing the relative contributions of nature (genetics) and nurture (experience) to sexual orientation. Unfortunately, the current controversy is more a matter of scientific research versus religious doctrine. Fundamentalist Christians and Muslims insist that sexual orientation is entirely a matter of choice and can be reversed by "conversion therapy" programs. Rational scientists, on the other hand, understand the research data that show significant genetic determinants to sexual orientation. That doesn't mean that there are specific genes exclusively for homosexuality, but it does imply that we all have different constellations of genes that predispose us to our particular sexual orientation.

A little genetic information may be useful to the reader. First let us dismiss the pseudoscientific argument that homosexuality can't have a genetic basis because if it did, it would go extinct. That's nonsense—there are lots of evolutionarily neutral traits— and even some maladaptive ones—that persist in the human genome. Furthermore, we don't know that homosexuality is maladaptive. Sure, gays usually don't sire children, but they may enhance the fitness of their kinship group through their actions. Joining a celibate priesthood would also seem maladaptive by that argument, yet we know that historically many priests have acted to enhance the status, wealth and ultimate reproductive success of their kinship group. Second, we should note that pedigree analysis demonstrates that male homosexuality is common in certain families and not in others. Similarly, female homosexuality is also common to certain families (but not normally the same families as male homosexuals). Third, male twin studies show that if one monozygotic twin is gay, the probability of the other (genetically identical) twin being gay is about 50 percent. If one dizygotic (fraternal) twin is gay, the probability that the other is gay is about 22 percent. If one brother is gay the probability of his only brother being gay is also 22 percent (fraternal twins are genetically the same as non-twin siblings). Finally, the probability that the adoptive infant sibling of a gay child is also gay is only about four percent. That is the same as the incidence of homosexuality in the general population. These simple data show a strong, but certainly not total, genetic influence on the trait. Furthermore, the data suggest that "re-education" programs to convert homosexuals to heterosexuals will generally not work. As any gay person will tell you, it is never a simple matter of "choice"—their genetics (or their God!) made them that way. The consensus of behavioral and social scientists and medical and mental health professionals is that homosexuality is a normal variation in the spectrum of human sexuality. The problem is not with these people's sexual orientation, but rather with the religionists who seek to impose their archaic and mistaken beliefs on them. Surveys vary, but about four percent of Americans self-identify as either lesbian, gay, bisexual or transsexual (LGBT). In some locations the frequency is higher: San Francisco has 6.2 percent LGBT; Portland has 5.4 percent; Austin has 5.3 percent; and Seattle has 4.8 percent. Some of the lowest rates are seen in the religious South or Rust-Belt communities where LGBTs may conceal their sexual orientation to avoid social discrimination and physical risk—in Birmingham, AL the reported rate is 2.6 percent; in Pittsburgh it is 3.0 percent; in Memphis it is 3.1 percent and in Milwaukee, 3.5 percent (Leonhardt and Miller. "The Metro Areas With the Largest, and Smallest, Gay Populations." *NY Times* 3/20/2015).

An approximation of overall LGBT frequency in the US would be about 4.0 percent. Thus, there are at least 8.8 million LGBTs among the adult (age 15–45) population. These 8.8 million or more of our fellow Americans may be subjected to the most cruel, humiliating and physically dangerous abuse.

They may be rejected by their families, humiliated by their schoolmates, fired by their employers, and denied full membership in their churches. The US Department of Health and Human Services estimates that gays and lesbians commit *"as many as 30% of completed youth suicides each year."* The religiously based homophobic bullying that precipitates many of these suicides is finally being recognized as a national disgrace!

The late Fred Phelps, former lawyer and Founding Pastor of the Westboro Baptist Church in Topeka, KS, exemplified the extreme religious homophobes who embarrass this nation. Phelps and his band of pathetic "Christians" staged demonstrations at the funerals of fallen soldiers and Marines to protest our national acceptance of homosexuality and our policy of allowing gays to serve in the military. They are the ones at funerals who held up the placards reading, *"God Hates Fags," "Thank God for 9/11," "America is Doomed," "God Hates You," "Thank God for Dead Soldiers,"* etc. They are certain that *"God Hates America"* because some of us tolerate *"the wickedness of homosexuality."* These people may be fringe extremists but they constitute a real concern for grieving families—they recently picketed a funeral in Clarence, NY and threatened to appear at a funeral for a young Marine hero in Jamestown, NY. To be entirely accurate we should emphasize that Phelps' flock is comprised almost entirely of his family members. All of his 16 children were indoctrinated into his "faith" as youngsters. Most managed to find similarly inclined Christians to marry and thereby "breed" more extremists, but three managed to escape their father's control. Nate Phelps, the seventh child, failed to buy into his father's *"extreme version of Calvinism"* and now lives in Calgary where he works for the Center for Inquiry, a global, nonprofit, secular organization that promotes science, reason and humanist values. In his blog, Nate describes a childhood of *"extreme physical punishments and abuse, extreme dietary and health requirements, and other extreme expectations."*. He has publicly spoken out against his father's beliefs and has become an advocate for LGBT causes. Fred Phelps' hateful rhetoric has turned at least one of his children to atheism, and has probably driven untold numbers of "Christians" to reexamine their beliefs. Phelps died March 19, 2014, leaving a sad legacy.

The US Supreme Court determined that the hateful actions of Westboro Baptist Church members at military funerals were protected under the First Amendment. I must concur with the Court decision—I would like to see the Church's behavior curbed, but not at the cost of compromising the First Amendment.

* * *

Before ending this rant I am compelled to mention the blatant sexual abuse and discrimination that permeates Abrahamic tradition. In Numbers (31:13) we read that Moses instructed his victorious army commanders to *"kill every male among the little ones, and kill every woman that hath known a man by lying with him; but all the women-children that have not known a man by lying with him, keep alive for Yourselves."* Today "civilized societies" don't force women into overt sexual slavery, but we still deny them equal rights. Orthodox Jews still practice a rigorous double standard that is rooted in their scripture. "Modesty buses" in downtown Jerusalem require women to wait until all the men have boarded and then sit or stand in the rear. *Di Tzeitung*, a NYC orthodox newspaper, removed Hilary Clinton's image from the group photo taken in the White House Situation Room during the assault on Bin Laden's Pakistan compound. Comments by Shlomo Rosenstein, a city councilor in Jerusalem, explain both the "modesty buses" and the unauthorized "photoshopping" of Secretary Clinton: *"This really is about positive discrimination, in women's favour. Our religion says there should be no public contact between men and women, this modesty barrier must not be broken."* In Beit Shemesh, a suburb of Jerusalem, the "modesty barrier" is more extreme. Recent accounts tell of Ultra-Orthodox (Haredi) men harassing schoolgirls whose attire doesn't meet their conservative standards of modesty. Little girls (as young as eight) dressed in school uniforms (long skirts and high-necked, long-sleeved shirts) are taunted, spit on and called whores as they walk to their Orthodox all-girls' school. Parents escorting their children are subjected to the same treatment. The Haredim insist that the gender integration insults their faith, but in reality, they are just seeking to perpetuate their religion's longstanding subjugation of women.

The Catholic Church also claims that it treats the sexes equitably, but persists in disallowing women to serve as priests. As recently as July 2010 the Vatican reaffirmed its position and stated that anyone *"attempting ordination"* of women would be automatically excommunicated. The punishment for this *"grave sin"* was included in the same category as the revised procedures for dealing with pedophiles. The Church was essentially saying that the two crimes—ordination of women and pedophilia—warranted equivalent punishment. Considering that the Church has almost never punished known pedophiles yet has reputedly excommunicated more than 100 female ordained priests, it hardly seems equivalent. A church that knowingly has ordained thousands of pedophiles and tens of thousands of homosexuals shouldn't be afraid of a few women.[30]

* * *

I can't write about oppression without some mention of slavery. I have recently finished reading the *Slave Narratives*, a collection of essays by escaped slaves who later lectured on the American abolitionist speaking circuit. A common observation of many of these former slaves was the remarkable piety amongst the cruelest slaveholders. If you read Exodus and Leviticus you will understand why this was so—slavery is fully endorsed in the Hebrew Bible. Examples abound, but consider: *"You may also purchase the children of such resident foreigners, including those who have been born in your land. You may treat them as your property, passing them on to your children as a permanent inheritance* (Leviticus 25:45). So too, with the Christian Bible—in St. Paul's Ephesians (6:5) you will read: *"Slaves, obey your earthly masters with deep respect and fear. Serve them sincerely as you would serve Christ,"* and in Timothy (6:1-2) you will find: *"Christians who are slaves should give their masters full respect so that the name of God and his teaching will not be shamed. If your master is a Christian, that is no excuse for being disrespectful. You should work all the harder because you are helping another believer by your efforts."* Devoutly Christian slaveholders rationalized their behavior—I daresay many, if not most, believed they we doing "God's work."

* * *

Thus, Christianity provided the moral justification for slavery in America and, by extension, the war that would ultimately claim 750,000 American lives. Confederate Vice President Alexander Stephens proudly proclaimed that the Confederacy would be

> *the first government ever instituted upon the principles in strict conformity to nature, and the ordination of Providence, in furnishing the materials of human society. … With us, all of the white race, however high or low, rich or poor, are equal in the eye of the law. Not so with the negro. Subordination is his place. He, by nature, or by the curse against Canaan, is fitted for that condition which he occupies in our system. The architect, in the construction of buildings, lays the foundation with the proper material—the granite; then comes the brick or the marble. The substratum of our society is made of the material fitted by nature for it, and by experience we know that it is best, not only for the superior, but for the inferior race, that it should be so.*

> *It is, indeed, in conformity with the ordinance of the Creator. It is not for us to inquire into the wisdom of His ordinances, or to question them. For His own purposes, He has made one race to differ from another, as He has made "one star to differ from another star in glory." The great objects of humanity are best attained when there is conformity to His laws and decrees, in the formation of governments as well as in all things else. Our confederacy is founded upon principles in strict conformity with these laws.*

The legacy of slavery is found in the racism that persists to this day. The Civil War did little to change the thinking of many southern Christians—blacks were still seen as sub-human animals. Between 1877 and 1950, nearly 4,000 black citizens were lynched in our 12 most "Southern" states. In his infamous 1963 "Segregation Now" inaugural speech, Alabama Governor

George Wallace defiantly opposed the federal mandates for integration "in God's name," and likened the federal action to that of an anti-Christ. While we may have integrated schools, racial inequality is still a serious problem in this country. It is the legacy of religiously endorsed slavery!

While slavery has declined in most of the world, there are places where it still prevails. Mauritania is located in West Africa—14–26 degrees north of the equator. It is mostly desert and home to about 3.4 million people. In 2012 it was estimated that 10–20 percent of the population (340–680,000 people) lived in slavery. More conservative estimates in 2014 indicate that about four percent of the population (~156,000 people) are still held as slaves. Mauritanian slaves are descendants of black Africans and are "owned" mostly by lighter-skinned Beydanes, the descendants of Berbers. Although slavery was made illegal in 2007, the law is very difficult to enforce. Slavery is central to the culture and fully endorsed by the overwhelming Sunni Muslim majority. Most slaves accept their fate as part of a Divine natural order. Anyone who questions slavery is seen as anti-Islamic. (It is worth noting that Mauritania is one of those few nations that still has a death penalty for atheism.)

***

Of course ultimately, it is often the most religious who are the most oppressed—but they often fail to recognize it. If you told any very religious person that they are oppressed they would think you are crazy—they don't consider themselves oppressed or manipulated because they have never given it any serious thought. Karl Marx's famous maxim that *"Religion is the opium of the people"* says essentially that—religions so anesthetize people that they are unable to see that they are being manipulated and controlled. Historically, most Christians have been indoctrinated to believe that they should humbly accept their status, believing that in the end they will be rewarded—their god will ultimately settle all injustices for them.[31] As children, most Christians learn the eight Beatitudes from Christ's Sermon on the Mount. The first of these, according to Matthew (5:3-10), was: *"Blessed are the poor in spirit, for theirs is the kingdom of heaven."* While debate continues about the meaning of *"poor in spirit,"* the important thing is that Christians have traditionally been taught that it meant materially poor, not spiritually poor. Luke (6:24) omits the *"in spirit"* qualifier and records Christ's utterance simply as, *"Blessed are you who are poor, for yours is the kingdom of God."* Luke clearly meant "poor" as in "lacking material possessions." In various other places, Luke praises the poor and condemns the rich, so there should be little doubt of his intent (e.g. *"But woe unto you that are rich! For ye have received your consolation"* [Luke 6:24]). Recall also Matthew 19:24—*"It is easier for a camel to go through the* eye *of a needle, than for a rich man to enter into the kingdom of God."* The remaining Beatitudes variously bless the meek, the persecuted, the justice-seekers, the peacemakers, the merciful, etc. Collectively, these verses encourage the downtrodden to submit to their miserable lives with the assurance that their poverty is a blessing and that they will be compensated for their suffering in heaven. They know with smug certainty that they will have an easier time getting into heaven than their rich, unjust and powerful oppressors. Remarkably similar sayings can be found in the Dead Sea Scrolls, written 70–130 years before the birth of Christ. It would appear that this form of religious psychological manipulation has been around for more than two millennia.

Encouraging the poor, the downtrodden, and the enslaved to submit to their miserable lives and not question authority implies that the clergy has traditionally collaborated with other rich and powerful oppressors. Religious submission predisposes people to oppression. The secure belief in a rewarding afterlife serves the oppressor at least as much as it comforts the oppressed.

This is what Napoleon meant when he said, *"Religion is what keeps the poor from murdering the rich"*—religions so mask reality that the masses will often endure their miserable status without rebelling.

In Defense of Religion

"We must respect the other fellow's religion, but only in the sense and to the extent that we respect his theory that his wife is beautiful and his children smart." —H. L. Mencken

"This would be the best of all possible worlds, if there were no religion in it." —John Adams

"I cherish everyone's right to their religious beliefs, no matter how comical." —Herman Melville

"Religion is all bunk." —Thomas A. Edison

Despite my strong inclination to follow Edison's simplistic assessment, in fairness to my few religious friends, I must consider the arguments for the unique contributions of religions to human welfare. By unique, I mean those contributions that are not performed just as well or better by equivalent non-religious institutions. After you sort out all the obviously false, supernatural, religious and inane claims, there are only a few contributions that are possibly unique to religion. Let me first identify three that are widely considered unique but are not: spiritual growth, moral development, and happiness.

One. Spiritual growth is not an exclusive benefit of religion. In fact, I have argued earlier that in many cases religion impedes personal spiritual growth by prescribing beliefs, practices and outcomes. The most controlling religions appear to prevent the development of anything resembling genuine individual spirituality.

Two. Moral development, likewise, is not a benefit unique to religion. Humans are born with an innate capacity to act morally, and most secular societies reinforce that behavior just as well or better than religious societies. Religions often substitute ridiculous moral strictures that corrupt the person's innate moral compass. Even worse, many religions allow bad people to feel good about themselves by "forgiving" them if they make minor symbolic moral restitution. Yes, religions can promote virtuous behavior, but many other social institutions (e.g., families, schools, youth organizations and society in general) can and often do that just as well or better. Most modern secular cultures maintain social order and moral behavior without resorting to supernatural beliefs—and they appear to do it better than very religious cultures. Previously, we have argued that the moral progress of humankind has been impeded throughout history by the constraining forces of formal religions.

Three. Several recent surveys claim to show that religious people are happier than non-religious people and that happiness leads to better mental and physical health. Careful analysis of the available data negates virtually all of those assertions. We cited and displayed some of Paul Zuckerman's (2005) data in national "religiosity rates" elsewhere (Note **7**). If we compare those values with the "happiness rating" for various countries in the 2005–2009 Gallup World Poll we will see a profound negative correlation between religiosity and happiness—nations that consistently appear in the top 12 on the list of "Happiest Countries" (e.g., Denmark, Norway, Sweden, Finland, Australia, New Zealand, Canada, Austria and Switzerland) are among the least religious (have the highest percentage of non-believers). Consistent with that pattern, the very religious U.S. rarely appears in that "happy group"— the U.S. ranks 23rd among 178 countries in the 2006 Business World survey and 16th among 80 nations in the 2008 World Values Survey. Contrary to what you may have read, the available data show a *negative* correlation between religiosity and happiness among the industrialized nations of the world!

At the individual level a plethora of poor studies attempt to show a positive association between church attendance and personal happiness. One such study by the Austin Institute for the Study of Family and Culture was reported in Breitbart News Network (Dec 24, 2014) with the headline, "Religious People Much Happier Than Others, New Study Shows." It also appeared in the *Daily Mail* (UK), The *Telegraph* (UK) and some even less reputable news outlets. Without troubling you with details, the study found that people who attend religious services on a weekly basis are more likely to describe themselves as "very happy" than people who never attend. This research did not qualify for publication in a peer-reviewed journal but was widely cited nonetheless. More than fifteen years ago Diener and Seligman (2002. *Psychological Science* 13, 81–84) convincingly showed that when we statistically control for social relationships the association between church attendance and happiness disappears. Religious people who attend church regularly tend to have more social ties and if you take this into account, religion by itself does not predict or promote happiness.

A more recent study that examined regular church attendance in relation to personal happiness provides insight into how solid research results can be misunderstood. The study by Robert Putnam and Chaeyoon Lim (*American Sociological Review*, Dec. 2010) was widely publicized and misinterpreted by the popular press. Their results suggested a relationship between church attendance and individual happiness among American Christians but they took care to analyze their data completely. They determined that it is friendships within the congregation and not attendance *per se* that promoted individual happiness—a finding entirely consistent with Diener and Seligman's findings eight years earlier. According to one of the authors (Lim): *"Our evidence shows that it is not really going to church and listening to sermons or praying that makes people happier, but making church-based friends and building social networks there"*. The researchers noted that the *"spiritual aspects of religion"* did little to further a sense of well-being. Participants in the study who *"personally experience the presence of God"* and/or *"personally feel God's love in life"* reported no greater sense of well-being than people who do not believe. Confident non-believers exhibit life satisfaction, sense of well-being and happiness equal to those who confidently believe in God. It was the unfortunate ambivalent souls who suffered. In Putnam and Lim's study, the people who attended church but had no friends in the congregation did not experience any "happiness effect"—in fact, they were less happy than people that never attended church! We should note that their study also demonstrated that private religious practices (*e.g.*, praying and holding services at home) were not associated with greater life satisfaction.

A number of other recent studies purporting to show that individual religiosity promotes *personal* happiness, have inappropriately compared the *"most religiously committed"* to the *"least religiously committed,"* and naïvely included atheists and agnostics in the latter category. However, if you separate out *"committed agnostics/atheists"* from other *"least religiously committed"* and compare just them to the *"most religiously committed,"* the reported differences disappear. Both the *"committed agnostics/atheists"* and the *"most religiously committed"* were happier and more satisfied with life than the remaining *"least religiously committed"* or *"doubters."* It is ambivalence and doubt that correlates with unhappiness, not a lack of religious belief. The people who are least happy and most stressed are the conflicted individuals, many of whom have been religiously indoctrinated at one time, but later, by virtue of their experiences or intellect, come to question their faith. Many Americans fall into this category. If there is a principal cause for their unhappiness and pain, it might be their original indoctrination, not their lack of faith. People who have a confident worldview, whether religious or atheistic, seem to enjoy greater happiness and mental/physical health in their lives.

The frequent claims that religious people are healthier and therefore live longer than atheists are not supported by any evidence. In fact, nearly all of the countries listed above as having the highest "happiness scores" correlated to high rates of agnosticism and atheism also have the longest lifespans—the Japanese, who are 64–65% non-believers enjoy the longest lifespan (82.3 years) of industrialized nations. They are followed closely by Australia (81.8 years), Canada (81.4 years), Sweden (81.1), Switzerland (81.1), New Zealand (80.6), Norway (80.2), Germany (80.1) and the UK (80.1). All these nations have higher than average percentages of non-believers, and yet their people live the longest. In a list of 223 nations, the very religious US (only 3–10% non-believers), ranks 50th in longevity (78.4 years). Do you think there is a connection? I do. I would urge you to ignore all the anecdotal nonsense you hear about the healthful benefits of faith. (Of course I should also urge you to ignore all the political propaganda you've heard about the "inferior Canadian health care system"—if it's so bad why does the average Canadian live a full three years longer than the average American? The beer?)

Certainly there are many cases where religious individuals have experienced a healing effect from their faith—my grandmother Fuller's "holy water" worked wonders for my mother and certain of my siblings. But it is no different than any other placebo effect such as the many produced by "modern-day" over-the-counter herbals and nostrums, or the primitive incantations of a tribal shaman. If you believe it is helping you, it probably will. Religious placebo is no different or better than any other placebo and certainly doesn't constitute a unique quality of religion.

Finally, I can't resist noting the recent studies (*Time*, March 24, 2011) that have shown that young people in the US who attend church regularly are more than twice as likely to become obese in middle age than those who don't attend. Of course this again is just correlation, and the real causes are unclear. Still, I'm compelled to recall the bumper sticker that proclaims: *"If God had intended for me to go to church he would have given me a bigger ass to sit on and a smaller brain to think with."*

* * *

So are there any serious (unique) benefits to religion? Perhaps, but they may not outweigh their cultural side effects. I know just three worthy of consideration.

One. Religions may produce mental solace and help alleviate our fear of death. Ernest Becker in his book *Denial of Death* argued that a universal (primal) human anxiety arises from our terror of death. Becker saw us as a unique animal—one that can contemplate the great mysteries of the universe while occupying a physical body destined to be eaten by worms. That existential dilemma, and our awareness of our own mortality and ultimate death, forced us to invent religious illusions to shelter us from that "terrifying truth." Becker said that our need to control our primal anxiety is the basic motivation for much of our behavior and that our social structures, and most especially our religion-based cultures, serve principally to shield us from our awareness of death. At a personal level, this "denial of death" protects us from potentially debilitating terror and enables our daily functioning. Becker suggested that each individual's character is shaped by his or her particular devices for denying death. He further asserts that these same protective devices prevent us from acquiring genuine self-knowledge (and presumably, "authentic spirituality," as I have argued previously).

But more importantly, Becker believed that much of the evil in the world is a direct result of the behaviors and institutions that arise from our need to deny death. When members of a culture enthusiastically embrace a common death-denying belief system, any questioning of that system threatens the group's emotional security and self-esteem. If you fervently believe that you have earned eternal life as a result of your obedience to a particular faith and someone like Edison (above) tells you it is *"bunk,"* your security is threatened and you may become very angry and even violent. Becker reminds us that threats like this at the group level can and do lead to wars.[32] Of course, the hatred that many believers express for atheists is simply an aspect of their "worldview defense behavior"—atheists by their very existence are a direct threat to believers' emotional security.

Recently there has been a resurgence of interest in Becker's work through the emerging sub-discipline of "Terror Management Theory" (TMT). TMT has sought to directly test and apply Becker's work. The curious reader might want to watch the 2003 award-winning documentary *Flight from Death: The Quest for Immortality.* Reviewers have described it as a *"mind-blowing," "life-transformational film."* I found it to be a very useful summary of Becker's ideas and Terror Management Theory research. Anyone who expects to die one day should probably watch it! Seriously, it is hardly "mind-blowing" to the educated person. There are many other current and simpler descriptions of TMT that are accessible on the Internet—I would recommend any of the video clips of Sheldon Solomon's entertaining lectures on the topic at Skidmore College and elsewhere.[33]

We non-believers never really come to deny death; we just develop intellectual devices to help us ignore it most of the time.

Research suggests that those of us who come to understand ourselves as the physical and chemical products of organic evolution don't experience the same degree of terror that the unconsoled believer may feel. And, of course, we non-believers don't incur the same emotional indebtedness that the believer develops to his/her creator. It is that emotional debt that causes so many believers to act violently toward any challenge to their faith, but most especially toward any challenge to the reality of their god. In contrast, we non-believers rarely "hate" those who disagree with us! You may justifiably argue that science is simply my strategy for dealing with the terror of death. I won't argue with that. That's what Goethe was suggesting 300 years ago when he said: *"He who possesses art and science has religion; he who does not possess them, needs religion."* My knowledge of science affords me great spiritual security.

* * *

I do feel genuine compassion for those poor souls throughout history who were fully indoctrinated into an Abrahamic "Lake of Fire" mythology. Biological death can be a frightening enough passage without all the terrifying post-death tortures of Scripture. These people have to unfailingly adhere to their beliefs to protect themselves from unspeakable terror. The sad irony is that they cling to those belief systems for protection against the horrors that those very same belief systems instilled in them to control them. Fortunately, few modern Christian denominations still teach that depiction of hell, and the Muslims, of course, reserve that horrific torture chamber version of hell for the *"kafirs"* (unbelievers and non-Muslims).

In some ways, contemporary evangelical Christianity has made it too easy for its subscribers—they only need to claim Jesus as their personal savior to be born again, free of sin, with the promise of eternal life. If they can fully embrace that belief without question, it will serve them as a wonderfully effective death-denying, immortality illusion. They will never have to fear death.[34] Do I think people who are able to really deny death through that form of self-delusion are happier than me? You might as well ask me if I believe that *"ignorance is bliss."* It is quite possible that such people are happier but only in a very superficial sense—they must deny themselves real knowledge of the universe and what I consider to be authentic spirituality. I'm free to find intellectual fulfillment to a degree they cannot (and might not dare, if they were able). In the absence of faith I can develop a conscious and rational understanding of death that isn't all that frightening. Disappointing, yes; but frightening, no! A recent comprehensive research study at Oxford (Dr. Jonathan Jong *et al.*, "The religious correlates of death anxiety: a systematic review and meta-analysis." *Religion, Brain and Behavior* 1–17, Mar. 2017) seems to support my position. When death anxiety levels were compared, atheists generally showed the least fear of death. Individuals who sought religion for social and emotional support ("extrinsic religiosity") exhibited the greatest fear of death ("death anxiety"), whereas individuals who embraced religion through "true belief" ("intrinsic religiosity") exhibited a lower fear of death comparable to that of atheists.

Thus, I will concede that the emotional solace of religious death-denial may be a positive attribute for many devout individuals facing death, but is no greater than the comparable solace enjoyed by atheists.

* * *

There is a profoundly negative side to death-denying worldviews that warrants repeating—when a single immortality scheme (i.e., religion) dominates a culture, country or region, bad things almost always happen. As Sheldon Solomon suggests, it often comes down to: *"My god is better than your god and we'll kick your ass to prove it."* It is no exaggeration to say that nearly all the major human conflicts over the past two millennia have been due to differing worldviews and death-denying strategies.[35]

So, do I think that the death-denying benefit of religion for individuals outweighs the loss of hundreds of millions of lives from worldview conflict in the 20th century alone? Certainly not![36] As adults we should learn to embrace death as a natural process. I'm convinced that an appreciation of the certainty and finality of death endows non-believers with a greater sense of humanity, empathy and rational morality than believers. Ultimately, we must find ways for all humans to manage their death anxiety through creative activities and enjoy the enhanced self-esteem that positive action can engender. The world religions could take the lead in this goal by promoting tolerance and teaching their followers to find personal fulfillment through altruistic action and not blind faith. Of course there are some minor religious sects today that actively and consistently practice non-violence and service, but they are a rarity. Frankly, I'm not optimistic about our future—as a species we have the ability to bring about our own extinction, and I'm afraid that is where excessive religiosity will eventually lead us.

Two. A second possible benefit of religions is their capacity for community-building. By indoctrinating followers into a common belief system and set of core values, religions could create a unity that promotes group cooperation. Some people have theorized that religion itself evolved because it enhanced group cooperation—individuals who shared core values would cooperate more and would have enhanced evolutionary fitness. As noted earlier, there is a considerable body of research showing that cooperative behavior in humans is an evolved trait. For example, it is no coincidence that humans are the only primates that have white sclera (the area around the colored iris) in their eyes. The white sclera makes it easy for others, even at a considerable distance, to determine the direction of our gaze. Psychologist Michael Tomasello argues that this uniquely human feature indicates that modern humans evolved within a cooperative social environment (*NY Times*, 13 Jan 2007). Gaze detection ability in conjunction with theory of mind would enable others to anticipate our actions with respect to identifying food sources or escaping danger. In a cooperative sharing society the trait would be quite advantageous, but in a group of selfish individuals it would be detrimental. (See note **38** below.)

It appears that most of our innate capacity for cooperation and community-building was in place long before the monotheistic religions took hold. Religions do, however, appear to selectively reinforce many of the pre-existing traits. Recent studies suggest that cooperation between individuals is enhanced when participants visibly display common beliefs (e.g., wear a crucifix or a Star of David pendant). William Irons, Professor of Anthropology at Northwestern University, has expertise in the evolution of moral systems and has argued that humans exhibit a strong psychological propensity to learn culturally specific signals of commitment, and these signals facilitate within-group cooperation (just like street-gang "signs").[37] As noted elsewhere, shared belief systems and standards of fairness have enabled human economies to thrive and expand.

I am willing to concede that religions can promote cooperation between believers of a common faith—religions and religious leaders have been associated with many honorable social causes throughout history. Unfortunately, they have also been associated with an equal number of dishonorable actions. Furthermore, there is overwhelming evidence that any benefit resulting from community-building is largely offset by an increased hostility toward out-groups. We have discussed this problem adequately in (1) above.

Three. Religions probably promoted further evolution of prosocial and cooperative behavior as larger and more complex societies developed. Prosocial behaviors involve concern for the rights of others and include empathy, altruism and norms of fairness.[38] Typically, the various religions would like you to believe they are the sole source of these mostly innate behaviors. Of course, they are not. In fact, some recent research suggests that religiosity significantly inhibits the development of altruism in youngsters in diverse cultures.[39]

We will concede that religions have often provided the reinforcement that promoted a more widespread expression of prosocial behaviors within their own group. As noted previously, direct and indirect reciprocation and altruism are clearly genetically based products of our evolutionary history—religions are human inventions in response to underlying psychological needs. Religions change by cultural evolution, whereas most prosocial behaviors respond to both natural and artificial selection and undergo biological evolution. We acknowledge that during our recent evolutionary history religion may have acted as a selective agent that favored genes for prosocial behaviors. That is, a powerful religious culture will normally provide rewards (potential fitness gains) for within-group, prosocial behavior, and will punish anti-social behavior. Conversely, however, religions may punish out-group-directed prosocial behavior and reward out-group-directed antisocial behaviors. The real question is whether the net effect of religion on overall human prosocial behavior is positive. The jury is still out on that issue, but I would suspect that at best, it's a draw.

* * *

We have failed to conclusively answer the question of whether religions actually do more good than harm. I suspect their contribution is mostly negative. I will, however, suggest that religions have the potential to advance human society if and when they drop the supernatural nonsense and encourage practical and constructive activities that enable followers to earn self-esteem through *genuinely* altruistic acts. At the very least we should acknowledge author Robert Wright's admonition: "Any religion whose prerequisites for individual salvation don't conduce to the salvation of the whole world is a religion whose time has passed."

* * *

The sad irony remains that most religions today serve mainly to shelter us from the terrors they have generated within us. That is, they are good at alleviating some of the problems they cause. That's pretty depressing, isn't it?

A FINAL WORD

It seems that we are on the threshold of great breakthroughs in our understanding of human "belief." I would urge the curious and intelligent reader to stay abreast of the advances in our knowledge—as we gain a greater understanding of the brain and the manifestation we call "mind" we will, like Toto, pull back the curtain to fully reveal that the "Wizard" is not really a supernatural being but rather a biologically decipherable entity. Stay tuned!

Notes

1. When Galileo died in 1642 following nine years of house arrest, he was buried in an unconsecrated grave. Ninety-five years later, the Church softened its position somewhat and permitted his admirers to move his body to a tomb in Santa Croce Basilica in Florence (not far from Michelangelo's tomb). At that time relic seekers removed parts of Galileo's skeleton including three fingers, a vertebra and a tooth. Most of these parts were eventually misplaced and given up for lost, but the middle finger of his right hand ended up on display at Florence's Museum of the History of Science. Today that finger rests, rudely pointing skyward, within a beautiful gold-encrusted, egg-shaped glass goblet. Some of us prefer to think that the great scientist's spirit is "flipping off" the Church that has historically impeded scientific progress. (I briefly considered sub-titling this little book "Galileo's Middle Finger" but discovered that there already were several books and essays with that same title, all discussing religious obstruction of science!)

   Galileo's punishment of house arrest was mild in comparison to the fate of many others. His predecessor, the Catholic monk Giordano Bruno, suggested in 1584 that there were probably many other suns and *"innumerable earths revolve around these suns in a manner similar to the way the seven planets revolve around our sun. Living beings inhabit these worlds."* Of course, Bruno was labeled a heretic and after years of torture in prison he was ultimately burned at the stake in 1600. Bruno's writings remain on the Vatican's list of banned texts. To this day the Catholic Church argues that he was burned not for his views on astronomy but for his *"theological errors."* Whatever! When he was sentenced to death by fire Bruno told his Inquisitors: *"Perhaps you, my judges, pronounce this sentence against me with greater fear than I receive it."* Indeed, the Catholic Church will never outlive Giordano Bruno and Galileo Galilei.

2. Klaus Lackner of Columbia, who works on artificial carbon sequestration systems (which could one day help reverse global warming) and "clean fossil fuel technology" recently said:

"Technology in general and energy at its base ultimately define the carrying capacity of the Earth for humans." Apparently Lackner, like some other academic "deep thinkers," has never fully contemplated the global consequences of the "infinitely available" energy he and others appear to pursue. The Second Law of Thermodynamics suggests that we will ultimately be overwhelmed by the waste heat of our energy transformations as we attempt to accomplish greater energy production. Another contemporary optimist, Erle Ellis of University of Maryland, Baltimore County, argues that human history is a record of ever-more-efficient food production technology and suggests that such technological progress can and will continue indefinitely. In a *NY Times* Op/Ed piece (13 Sep 2013), Ellis confidently asserted that overpopulation is not a problem and that there is no such thing as human carrying capacity for the planet. Ellis's main argument is essentially that overpopulation will not happen because it never happened in the past! That is an incredibly stupid argument. Like Lackner, Ellis fails to consider all the unintended consequences of the technology required to feed ten or more billion people. If the human race does survive to the end of the 22nd century, which I tend to doubt, conditions will likely be deplorable by current standards.

3. When I read *The Little Book of Atheist Spirituality* (Viking Translation by Nancy Huston—Viking-Penguin Press, 2007) I was delighted to discover that the author, French philosopher André Compte-Sponville, has used that same term, "Christian atheist," to describe himself. I now realize that quite a number of other people identify with the term and that it has a Wikipedia entry. I had stopped using that label when I realized that most of the supposed "Christian values" I identified with were neither unique nor original to Christianity. To be entirely accurate we should note that the values of forgiveness and charity that are considered central to Christianity may have seemed new to Western culture, but were not original to Jesus—several Eastern religions and/or philosophies had promoted those values centuries earlier. We often use the term "Christian values" to denote those common core beliefs, but we should be more precise in our language—they are not uniquely Christian qualities and we only call them that by convention. As noted elsewhere, these qualities are more the product of our biology than our belief systems. Whatever their origins, my sense of "christian values" embodies just a few main principles involving tolerance, kindness, charity, love, forgiveness,

fairness, and not much else. I know many other good people who embrace these same values without resorting to any belief in the supernatural.

4. Now the LORD had said unto Abram, *"Get thee out of thy country, and from thy kindred, and from thy father's house, unto a land that I will shew thee:*

And I will make of thee a great nation, and I will bless thee, and make thy name great; and thou shalt be a blessing:

And I will bless them that bless thee, and curse him that curseth thee: and in thee shall all families of the earth be blessed" (Genesis 12:1–3).

He (God) took him outside and said, *"Look up at the heavens and count the stars— if indeed you can count them."* Then he said to him, *"So shall your offspring be."*

Abram believed the Lord, and he credited it to him as righteousness. He also said to him, *"I am the Lord, who brought you out of Ur of the Chaldeans to give you this land to take possession of it"* (Genesis 15:4–7).

"To your descendants I give this land, from the river of Egypt to the great river, the Euphrates—the land of the Kenites, Kenizzites, Kadmonites, Hittites, Perizzites, Rephaites, Amorites, Canaanites, Girgashites and Jebusites" (Genesis 15:18–21).

Few religious writings have caused more strife than these little pieces of Scripture. What was probably a bit of wishful thinking by an early Semitic author has generated endless discord in the Middle East.

5. Recent explorations deep within Bruniquel Cave in southwest France have revealed elaborate stone structures that may have served a ceremonial/religious function for Neanderthals 176,000 years ago! These are among the earliest hominin constructions ever found. (See *Nature* 25 May 2016; *The Guardian* 25 May 2016). We should further note that some records suggest that, at least in some cases, Neanderthals decorated their graves. Such burial practices by Neanderthals are not so remarkable if you consider that certain other great apes, some cetaceans, and elephants all appear to mourn the loss of family and respected friends. These species all use tools and have refined communication skills, sophisticated memory systems, some degree of self-awareness, and probably have dreams—they likely enjoy "transcendent mental experiences" much like humans. We know that they all have their distinct "cultures"; is

it far-fetched to suggest that some other sentient creatures may also have "religions"?

6. Some scholars suggest that higher cognitive functions in humans developed coincidental to language and that sophisticated religions and monotheism couldn't develop until these cognitive skills were in place. In his book *The Origin of Consciousness in the Breakdown of the Bicameral Mind* (1976, Houghton Mifflin, NY), Julian Jaynes (1920–1997) argued that ancient people did not experience consciousness and self-awareness as we know it. Instead, their behavior was directed by auditory hallucinations that they interpreted as *"the voices of their chief, king or the gods."* These "commanding voices" originated in the right frontal brain areas that correspond to the speech centers (Broca's and Wernicke's areas) of the left lobe. Jaynes' ideas were soundly criticized when he first presented them in the 1970s, but recent work lends support—specifically, neuroimaging studies show that auditory hallucinations arising in the right temporal-parietal lobe are transmitted directly to the left temporal-parietal lobe. In Jayne's primitive "bicameral" human mind the early human functioned much like a modern-day schizophrenic whose behavior is governed by auditory hallucinations. In response to sensory stimulation, the person would hallucinate a voice in the right brain and obey its commands without ever being conscious of the mental process. With the emergence of metaphorical language (and later, writing,) the bicameral mind was gradually replaced by modern consciousness wherein the subject is more aware of his mental states. It has been suggested that such an evolutionary advance could be accomplished in just a few thousand years. The "imaginary friends" of so many small children and the schizophrenic episodes of occasional adults remain in the human population as vestiges of this earlier form of "cognitive" behavior. In Jaynes' own words, *"In connection with the personal god, it is possible to suggest that a part of our innate bicameral heritage is the modern phenomenon of the 'imaginary' playmate. According to my own research as well as other data (Singer & Singer, 1984), it occurs in at least one-third of modern children between the ages of 2 and 5 years, and is believed now to involve very real verbal hallucinations. In the rare cases where the imaginary playmate lasts beyond the juvenile period, it too grows up with the child and begins telling him or her what to do in times of stress."*

Space doesn't permit me to expand further, but suffice it to say that after 30 years Jaynes' ideas are gaining more respectability among neurophysiologists, psychologists and philosophers. I had the good fortune of meeting Julian Jaynes when he was a convocation speaker at Fredonia many years ago. I have always been fascinated with the possibility that our first gods were likely auditory hallucinations in the bicameral brain! Jaynes' book is pretty heavy reading; the curious reader might gain more by consulting Michael Persinger's 1987 book, *Neuropsychological Bases of God Beliefs* (Praeger, NY) or John Geiger's 2009 book, *The Third Man Factor* (Perseus Books, NY).

7. Recent estimates give the following numbers for some of the world's major "religions": Christian = 33% (2.25 billion); Islam = 21% (1.45 billion); Non-religious = 16% (1.1 billion); Hindu = 14% (950 million); Buddhist = 6% (400 million); "Primal-indigenous" = 6% (400 million); Judaism = 0.22% (13.2 million). Note that both Hindus and "Primal-Indigenous" religions are polytheistic. In addition, the many Christian religions that embrace a Trinitarian belief and pray to their many saints that serve as "minor gods" are effectively polytheistic—they just don't acknowledge it! Also note that about one-sixth of the world's population subscribe to no religion at all and are mostly atheist/agnostic. Among the estimated 13.2 million Jews in the world, about 4.5 million (~ 34%) consider themselves secular and unaffiliated and probably should be counted among the "non-religious." In addition, the world's 400 million Buddhists fall somewhere between atheistic and polytheistic. They certainly don't embrace the Judeo-Christian "one god of the desert," and many Buddhists consider belief in a Supreme Being antithetical to their path to enlightenment. Some of them do venerate the disciples of the historical Buddha and have various minor meditational deities, but none to my knowledge have an interventionist personal "god" or "gods" in the Western sense.

The reader should be aware that the number of professed atheists/agnostics in the United States (~3–10%) is probably a gross underestimate due to social coercion, but is still lower than in other scientifically literate, developed nations. Recent data show the following minimum-maximum estimated percentages of non-believers (respondents indicating "agnostic," "atheist" or "do not believe in God" on surveys) in some developed countries: Sweden

46–85%; Denmark 43–80%; Norway 31–72%; Japan 64-65%; Finland 28-60%; France 43-54%; South Korea 30-52%; Germany 41-49%; Russia 24-48%; Netherlands 39-44%; Britain 31-44%; Belgium 42-43%; Canada 19-30%; Switzerland 17-27%; Australia 24-25%; Taiwan 24%; Italy 6-15%; Mongolia 9%; Portugal 4-9%; **USA 3-9%**; Albania 8%; Argentina 4-8%. You can see we are in good company with the intellectual/educational heavyweights of Mongolia, Portugal, Albania, and Argentina. (Data from: Zuckerman, P., 2005. "Atheism: Contemporary Rates and Patterns" in *The Cambridge Companion to Atheism*, ed. by Michael Martin, Cambridge University Press: Cambridge, UK).

The "US Religious Landscape Survey" conducted by the Pew Forum in 2007 gives a little more insight into American religious affiliations and beliefs. That study determined that 16.1% of Americans claim no religious affiliation. Those unaffiliated Americans included 1.6% atheists and 2.4% agnostics, and a third group identified as "Nothing in Particular" or NIPs. Among the NIPs, 6.3% were "secular unaffiliated" while 5.8% were "religious unaffiliated." Thus, the total of atheists, agnostics and secular unaffiliated ("non-believers") was 10.3% according to the 2007 Pew poll. More recent studies suggest the actual number of atheists/agnostics is closer to 26% (Gervais and Najle, 2017—see Foreword). Most studies conclude that the world's most rapidly growing religious category in the past two decades has been "nonbelievers." Thank God for that!

8. The Theodicy Problem is a philosophical conundrum presented by the god of Abrahamic tradition. It is often stated as a trilemma— there are three statements about god, and it is impossible to logically reconcile any two without implying that the third is false. The three statements are: *1. God is omnipotent. 2. God loves us. 3. Evil exists.* Try it yourself. It appears impossible to integrate all three into a coherent concept of god. The Greek philosopher Epicurus (341–270 BC) anticipated this problem when he wrote:

> *Is God willing to prevent evil, but not able? Then he is not omnipotent.*
> *Is he able, but not willing? Then he is malevolent.*
> *Is he both able and willing? Then whence cometh evil?*
> *Is he neither able nor willing? Then why call him God?"*

Walter Stephens, Charles S. Singleton Professor of Italian Studies at Johns Hopkins University, has proposed that the

Church invented witches and demons in order to explain the existence of evil in the world and thereby resolve the Theodicy Problem. In his words: *"I think Witches were a scapegoat for God."* Yuval Noah Harari offers an amusing alternative explanation—there is a single omnipotent God but He's evil. But few theists embrace that belief!

9. The majority of the founding fathers, especially in New England, were deists and almost none were "Christians" in the contemporary sense—in fact, many used that term derisively! They did not believe in the Divinity of Jesus, although some found merit in his teachings. Nor did they believe in a personal God in the conventional Christian sense. Their occasional use of the word "God" in their writings should be considered more metaphorical than literal—in much the same way contemporary atheistic scientists use it as a synonym for the unknown. The "God" of the deists was a noninterventionist creator who permitted the universe to run itself according to natural laws. They believed that reason was a gift from that creator that enabled them to understand their place and purpose in their natural world. Deists did not embrace myth and supernatural explanations, and their "god" was more a metaphor for the unknown ultimate causation than "an invisible friend in the sky." They were really much more similar to modern agnostics than to modern Christians. Thomas Paine, who may have done more for the American Revolutionary cause than many of his more celebrated peers, has been given short shrift by historians because he was so anti-religious. Ten years after his death, a concerted effort was made to convince the public that he had undergone a deathbed conversion to Christianity. Nonsense—in *Age of Reason*, Paine wrote: *"All national institutions of churches, whether Jewish, Christian, or Turkish, appear to me no other than human inventions set up to terrify and enslave mankind, and monopolize power and profit."* Does that sound like a "Christian" to you?

10. George Herbert Walker Bush and James Danforth Quayle may have constituted the least articulate and informed team ever to lead this nation. "Dan" Quayle's gaffes are legendary. He is best remembered for insistently correcting a sixth grade public school pupil for failing to spell the singular of the word "potato" with an "e" on the end. Some of his more profound utterances include:

"It isn't pollution that's harming the environment. It's the impurities in our air and water that are doing it."

"I stand by all the misstatements that I've made."

"Verbosity leads to unclear, inarticulate things."

"Public speaking is very easy."

His mentor, the forty-first president, George H. W. Bush, wasn't any better:

"It's no exaggeration to say that the undecideds could go one way or another."

"I have opinions of my own, strong opinions, but I don't always agree with them."

"I am not one who—who flamboyantly believes in throwing a lot of words around."

"Those are two hyporhetorical questions."

"I'm not the most articulate emotionalist."

"Fluency in English is something that I'm often not accused of."

Bush's incredible insensitivity was evident when he visited Auschwitz in 1989 and proclaimed: *"Boy, they were big on crematoriums, weren't they?"*

11. During a recent debate with neurophysiologist-atheist-author Sam Harris, celebrated author-evangelist Rick Warren said that if he didn't believe that *"Jesus was the risen Lord,"* he *"wouldn't waste any time being good."* I hope that was a rhetorical device; if not, Warren is more stupid (and possibly evil) than I realized!

12. Some of the most interesting research in this area has been by developmental psychologist Michael Tomasello, Co-director of the Max Planck Institute for Evolutionary Anthropology in Leipzig, Germany and author of *Why We Cooperate* (October 2009, MIT Press). Tomasello and his colleagues have convincingly shown that humans are born altruists and only later learn *"strategic self-interest."* Children appear to be naturally cooperative, and their "instinctive" desire to help others is only later altered by their culture. Tomasello explains that all human children develop *"shared intentionality,"* a sense of what others expect to happen, and from this, a sense of group unity (e.g., the establishment of group norms and sense of fairness). Shared intentionality probably evolved very early in human history and facilitated cooperation in food-gathering and other social activities.

Recent studies on the evolution of the human sense of fairness are equally revealing. In an economic "ultimatum game" experiment, two players, a proposer and a responder, must divide a reward. It could be food or money or just about anything they value. The proposer offers a share to the responder. It could be anything from a 1% to a 99% share. If the responder rejects the offer as too low (unfair), both players receive nothing. When chimpanzees play the game with raisins, the responders rarely reject even the stingiest offers. Humans, on the other hand, almost always will reject any offer less than 20%. That is, they are willing to pay a price in order to punish fellow humans who fail to act fairly. There is no doubt—our "sense of fairness" (dislike of inequity) is an innate (evolved) human quality. Using the established model for estimating heritability, Bjorn Wallace at the Stockholm School of Economics showed a very high correlation between what one member of an identical twin pair proposed and what the other would accept. That is, their measures of fairness were similar—no such correlations were seen between fraternal twins. The high correlation between the fairness measures of identical twins indicates a genetic (evolutionary) basis to the behavior. We must point out that despite high heritability, an innate dislike of inequity may be modified by environmental circumstances in our species. (See: Maria Konnikova, "How We Learn Fairness." *The New Yorker*, Jan 7, 2016.)

At this point the reader might benefit from a more complete understanding of natural selection. *Sexual selection* was originally described by Charles Darwin as the *"special case"* where natural selection acts through an organism's ability to secure a mate and successfully reproduce. In many vertebrate species, the females select males on the basis of their extravagant displays and/or coloration. Males that are able to invest the most resources in their courtship displays are likely to be more "fit." More subtle displays of "fitness" may include acts of generosity, altruism and fairness. The individual who is able to be generous, altruistic and/or fair in interactions with others must be capable of securing more resources and thus would be a more "fit" (desirable) mate. Ancient humans likely preferred mates who performed acts of generosity, altruism and fairness over those that didn't—e.g., the successful hunter who shared fairly and generously was a more desirable mate than a successful hunter who didn't share fairly. Sexual selection via mate choice is an enormously important factor in the evolution of highly social species such as ours.

13. It is the height of arrogance and ignorance that certain Christians claim that *"the ethic of reciprocity"* is the *"Golden Rule of Christianity"* as if they had invented it. The rudiments of the "Golden Rule" are found in antiquity. A Late Period Egyptian (ca. 500 BC) papyrus contains the message, *"That which you hate to be done to you, do not do to another."* Confucian documents from the same period state, *"Do not do to others that which we do not want them to do to us."* Some insist that the earliest explication of the rule can be found in Leviticus (19:18) that urges us to *"Do unto others as you would have them do unto you."* The fundamental problem with these ancient versions of the Golden Rule is determining who are the *"others"* toward whom you are to act charitably. Are they family? More distant kin? Members of your tribe? Your church? Your nation? In the case of Leviticus 19:18 we know from reading the entire text that the act of kindness is reserved for kin only. Kin-restricted kindness is common in the Old Testament. For example, Deuteronomy 14:21 instructs: *"Ye shall not eat of anything that dieth of itself: thou shalt give it unto the stranger that is in thy gates, that he may eat it; or thou mayest sell it unto an alien: for thou art a holy people unto the LORD thy God."* That is, you and your "holy" family should not eat meat that is likely spoiled, but you may give it to a stranger, presumably poorer than you, who is a member of your tribe. But even better, you should sell it to the alien who is outside your gates. In this case alien literally meant someone who was not a member of your tribe, preferably a non-Jew. Christ's simple golden message of *"Do unto others as you would have them do unto you"* (Luke 6:31) does not make it clear how far we should extend our charity beyond our kinship group, but his story of the Good Samaritan (Luke 10:25-37) does. In that parable Jesus told how a traveling Jew was beaten by robbers, and how subsequent passersby including a priest (rabbi) and a Levite (fellow Jew) declined to offer assistance. Finally a Samaritan came by and aided him. Samaritans and Jews generally did not associate with one another, but this Samaritan served as an example of how we should expand our meaning of "other" or "neighbor." On another occasion when he was questioned by the Pharisees about "the Law" (Matthew 22:36–40), Jesus advised that his second commandment would be that *"You shall love your neighbor as yourself"* (Romans 13:8–10). We agree with the many Biblical scholars who believe that Jesus

intended "neighbor" to mean not your literal neighbor, but all of humankind, as he sought to introduce a new paradigm to the West. Acting altruistically toward kin is an ancient, biologically innate trait, but routinely extending that charity to non-kin is a more recently developed human behavior. Regrettably, this expanded scope of charity seems to have gained little acceptance among fundamentalist Jews, Christians and Muslims.

14. "Mirror neurons" were first described in the 1980s as a new class of brain cells within the premotor cortex of rhesus macaques—these unique cells fired whenever the monkey performed a goal-oriented action or when he observed another individual performing the same action (*"he mentally mirrored the action"*). The first and simplest explanation was that mirror neurons facilitated "motor memory" in the monkey and enabled him to "learn" to mimic actions of the others. More recent studies provided evidence of a comparable mirror neuron system (MNS) in humans and have suggested that they may do far more than enable mimicry. In addition to enabling the observer to "understand" goal-directed behavior in both himself and others, it has been inferred that an MNS is the evolutionary basis of empathy. It is not unreasonable to suggest that empathy (and "Theory of Mind," or ToM) is derived at least in part from our ability to imitate or emulate the feelings of others. At first that may seem a stretch, but let's consider a few examples. Vilayanur S. Ramachandran, Distinguished Professor of Neuroscience at the University of California, San Diego ("Rama" to his friends and colleagues), is a major proponent of MNS and explains it using the example of "pain empathy." I will paraphrase him: Imagine someone pokes my thumb with a needle—nerve cells in my insular cortex fire and I will feel pain. When neurons in the adjacent anterior cingulate fire in response to the pain I process the experience as painful—the affective quality of the pain. It so happens that those same cells in the anterior cingulate that fire when I process my pain also fire when I observe *you* being poked by a needle. That's the physiological model of empathy and provides insight into how it evolved. But now take it to a higher level—imagine for a moment that you are a chimpanzee and you sample a new food source while others watch. It's terribly bitter so you naturally grimace and spit it out. Do you imagine the other chimps will go ahead and sample the same food and experience the same bitter sensation despite your example? Of course they won't—they have mirrored your

response and accordingly recognize the food as unpleasant without ever tasting it. Now imagine you are still that same chimp and if you observe your cousin, a *human*, react negatively to a food source—you will reject it as well. (Now if you switch back to your own human species, we hope you will have the good sense to cautiously avoid the food if you observe a chimp's negative reaction!) It is quite clear that both humans and chimps have a capacity to intuitively apprehend the mental and emotional state of others. That higher level "observational learning" is the basis for both empathy and ToM, and both appear to be facilitated by mirror neurons.

15. A surprising number of studies suggest that atheists, on average, are more moral than theists. We may not all agree on what is "moral," but I will begin by using some conventional metrics— crime rates, incarceration rates and divorce rates. First, we would note that recent research demonstrates that predominantly atheist countries have the lowest crime rates. Worldwide rates of "societal dysfunction," corruption, and violent crime are consistently lower in highly secular democracies than in "very religious America" (see: Gregory Paul, 2005. *Journal of Religion and Society*, Volume 7— online). And within the US, relatively secular regions like New England and the Pacific Northwest exhibit better "social functioning" than the Southern "Bible Belt." Homicide rates within the US are significantly higher in the pious South than in the secular Northeast. In fact, the rates of all major violent crimes (aggravated assault, robbery, forcible rape and murder) are all significantly higher in the "Bible Belt." Total violent crime rates for 2007 in relatively secular ME, NH and VT are 118, 137 and 124 per 100,000 population; total violent crime rate for relatively religious FL, LA and SC are 723, 730 and 788 respectively. Enough said.

Second, consider incarceration data—about 75% of all Americans claim to be god-fearing, but a slightly higher percentage of the nation's prison population claims to be god-fearing. On the other hand, more than 4% of Americans claim to be atheists, but only 0.2% of the nation's prison population are atheists. There are dozens of different surveys confirming this general pattern—disproportionately few nonbelievers, in this country and elsewhere, end up in prison. This is not new information—over 90 years ago, W.T. Root, Professor of

Psychology at the University of Pittsburgh, suggested, *"Indifference to religion, due to thought, strengthens character."* He had observed that *"Unitarians, agnostics, atheists and free-thinkers"* were *"essentially absent"* from prison populations.

Third, consider divorce rates—American Christians often argue that their moral superiority arises from their commitment to marriage and family. More than two decades ago, "Dr." Tom Ellis, then chairman of the Southern Baptist Convention's Council on the Family, claimed that proof of moral superiority could be seen in the extremely low divorce rate of only one out of 39,000 (0.00256%) among born-again Christian marriages. That marital success was so impressive that George Barna commissioned a study to confirm Ellis's remarkable claim. Barna and his wife, Nancy, had founded the Barna Research Group in 1988 as a marketing research firm to support Christian ministries. The findings of their first study shocked both Barna and the ministries that had eagerly awaited the results. Contrary to Ellis's claim, the 1999 study showed that the divorce rate among Baptists was actually 29%! The divorce rates by main religious categories were: Jews—30%; Born-again Christians—27%; Other Christians—24%; Atheists/Agnostics—21%. The highest divorce rates (34%) were seen among unaffiliated fundamentalist Evangelical Christians. Only Catholics and Lutherans had the same low divorce rate as atheists/agnostics. In 2008 Barna Research conducted a follow-up study with nearly the same results—Catholics and atheists had the lowest divorce rates and various fundamentalist groups had the highest. Fundamentalist Christians are still having a hard time rationalizing the data and prefer not to discuss it.

Christians commonly claim that atheists are less charitable and/or less honest than theists. Be very skeptical of any church studies that purport to show that "believers" are more charitable than "nonbelievers." Yes, you can poll people and find that about 50% of nonbelievers report giving to charity whereas 65% of "believers" claim to give. But most charitable giving by believers goes almost exclusively toward the support of their churches and missions and can hardly be considered philanthropy. It is a little harder to document atheist philanthropy because there are few formal atheist charities, and atheists are not identified as such when they give to secular charities. However, careful analysis of the data show that nonbelievers are every bit as charitable as believers; probably more so. It is worth noting that the top four philanthropists in the US in 2012—Warren Buffett, Bill Gates, George Soros and Mark Zuckerberg—are all avowed non-theists. Their combined philanthropy for 2012 was $5.12 billion. The combined lifetime giving of the first three alone exceeds $64 *billion*—that's a lot more than most churches have given in the same time frame. (The Mormon Church gives less that 1% of its annual income to charity.) Andrew Carnegie, certainly one of America's most famous and generous philanthropists, was an outspoken atheist.

In many behavioral studies assessing readiness to help others, it is atheists, not believers, who typically distinguish themselves. The classic 1953 study of YMCA associates by Murray Ross ("Religious Beliefs in Youths") showed that self-identified atheists and agnostics were consistently more willing to help the poor than were those who called themselves religious. Although the data are imperfect, there is indication that blood and organ donation rates among non-believers are greater than for believers. Blood and organ donations are clear examples of "unrestricted" charity—while evangelicals claim to be very generous, we know that over 80% of their "charity" is specifically directed to other evangelicals or potential converts. That's not charity; it's in-group restricted sharing! Most of us define charity as the unrestricted gift of money, goods or services to others based primarily on need.

There are a few direct experimental studies that purport to show that believers are more moral than nonbelievers, but their design is often flawed and their interpretations are dubious. Typically, subjects are asked to rate the seriousness of various "moral indiscretions" of others, and not surprisingly, the believers, knowing they are being watched by their "ceiling cat God," show themselves to be far more judgmental and less tolerant than nonbelievers. To suggest that intolerance and judgmentalism are valid measures of morality is ridiculous.

With respect to honesty, a series of papers since Franzblau's classic 1934 work ("Religious Belief and Character Among Jewish Adolescents," Columbia University Teachers College Contribution to Education, No. 634) have shown a significant negative correlation between religiosity and honesty. High levels of religiosity often endow the individual with a sense of moral superiority leading to what psychologists call "moral licensing." Their elevated self-image may liberate ("license") them to engage in unethical behaviors. Moral licensing is known to disinhibit selfish and dishonest behavior among both children and adults. At a global level, if we plot national measures for religiosity (e.g., frequency of praying according to the World Values Survey) against measures of national honesty (e.g., rate of corruption according to the National Corruption Index 2015) for countries throughout the world, we see a statistically significant negative relationship. More religion equates to more corruption. We all know that high levels of national corruption have profound effects on people's lives. Some recent studies have shown that individuals who believe in a moralizing ("punishing") god were more honest toward other members of their religion but not necessarily toward individuals of other religions.

With respect to honesty, a series of recent studies have revealed an interesting relationship between honesty/integrity and cursing (Feldman, G., Lian, H., Kosinski, K., Stillwell, D., "Frankly, We do give a damn: The relationship between profanity and honesty." *Social Psychological and Personality Science,* 1/15/2017. Many of us were taught that cursing is immoral and many religious people are careful to refrain from cursing. The Christian Bible has a lot to say about bad language. Yet these new studies found that individuals who use more swear words tended to have higher levels of verbal intelligence. More significantly, people prone to cursing exhibited higher levels of integrity and honesty than those who shunned cursing. To quote the authors: *"The consistent findings across the studies suggest that the positive relation between profanity and honesty is robust, and that the relationship found at the individual level indeed translates to the society level."* Perhaps we need to seriously reexamine what we consider moral and not moral.

Of course, it is quite possible that my definition of morality is different than yours. When I use the term "human morality" I am referring to "ethical behavior." Too many fundamentalist Christians and Muslims have limited understanding of morality, and as a result of their peculiar religious indoctrination believe that it deals principally with respect for their god and/or with sexual behavior! Test yourself—if someone is said to be "immoral," do you immediately think that they have violated sexual mores? Is a "hooker" more sinful than Bernie Madoff? How about the greedy bastards responsible for the world's most disastrous oil "spill" in the Gulf of Mexico—are they immoral? How about somebody who doesn't observe the same religious practices as you? Are atheists, by definition, immoral? I hope you answered "No," "No," "Yes," "No" and "No." If not, you should give it a little more thought.

When people fully embrace a religion they permit the "sacred laws" of that belief system to supplant their innate moral compasses. The loss of an innate moral compass and the development of "moral licensing" may allow them to perform all sorts of atrocities in the name of their faith. Adolph Hitler used his perverted version of Catholicism to rationalize his murder of Jews, gypsies, homosexuals, atheists and others. He was absolutely certain his actions were "moral."

16. "Cargo cults" are religious practices that have appeared when certain traditional tribal societies first interacted with technologically advanced cultures. Their rituals focus on acquiring the wealth ("cargo") possessed by the advanced cultures. The most famous cargo cults developed in New Guinea during WWII when native people observed military personnel unloading supplies from transport aircraft. They believed that they could also acquire cargo by building rude copies of the airstrips and aircraft that would cause their deities to deliver "cargo" intended for them. Cargo cults sprang up throughout New Guinea, Melanesia and Micronesia. One of the most famous and enduring of them is the John Frum cult that appeared in the late 1930s and is still active on the island of Tanna, Vanautu (formerly New Hebrides). According to one of the more credible accounts, a light-skinned native named Manehivi assumed the Western name "John Frum" and went from village to village on Tanna dressed in a Western-style white coat promising the treasures of Western culture including new clothes, abundant food and new homes. (One account suggests he appropriated the name from an American he met who he knew as "John from [i.e., "Frum"] America.") In John Frum's "New Age," all white people would leave the New Hebrides and the natives would gain access to all their wealth. John Frum's followers were required to reject all aspects of Western culture, including Christianity, and return to their native customs. In 1941 his followers united and left the Westernized villages, moved inland and resumed their traditional animist-based religion and polygamous culture. Some of the many churches seeking to "save" the natives depended on them as giving parishioners. And, of course, all of the copra (coconut oil) plantations depended on the natives as a source of cheap labor. When the natives left the

settlements to follow John Frum, the economy was destabilized. In an effort to "restore order," the European colonial authorities arrested and imprisoned John Frum. He and the leaders of his movement were publicly humiliated and exiled to another island, all with the blessings of the Christian churches and plantation owners. In spite of the attempts to suppress the movement, John Frum's followers increased in numbers. During WWII they built the now-famous symbolic landing strips to encourage aircraft to land and bring them Western "cargo." By the late 1950s a leader of the John Frum movement named Nakomaha founded the "Tanna Army," a non-violent, quasi-religious group that marched in parades wearing white T-shirts with "T-A USA" written on them. They are active to this day—each February 15, now known as "John Frum Day," they paint their faces in ritual colors, don their Tanna Army T-shirts and march military-style in anticipation of the return of their martyred leader. The year of his return is not known, but they are convinced it will be on February 15th some year soon. A few of my close relatives and friends celebrate John Frum Day with the same religious fervor that we celebrate Easter.

Nearer to home, the Rastafari movement that began in Jamaica in the early 1930s is a monotheistic Abrahamic spin-off that regards the late Ethiopian Emperor Haile Selassie (*aka* Ras Tafari Makonnen) as a divine prophet and treats his speeches as holy text. Rasta now has over a million followers worldwide. Part of its success may be related to the fact that it promotes the use of marijuana for spiritual enlightenment and has been linked to Bob Marley's reggae music.

17. Most of us have been indoctrinated to certain images of a "Christian Heaven" with gates, clouds, angels with harps, etc. My mother told us of such a wonderful place that excluded everyone except Catholics! She was only following the faith she learned from her sainted grandmother Mary Margaret Reardon Murphy Higgins. Prior to "Vatican II" (The Second Ecumenical Council of the Vatican—1962–1965), only good Catholics who were able to receive priestly absolution could enter Heaven. Vatican II changed that—good non-Catholics could go to Heaven, but being a Catholic was much better—as a Catholic you could be an asshole and still get into Heaven! Of course today, born-again Christians enjoy that same privilege. Church of Scientology founder L. Ron Hubbard's description of Heaven was more convincing than my mother's because he was able to actually recall details from one of

his visits there: *"The entering grounds are very well kept, laid out like Bush [sic] Gardens in Pasadena."* It sure sounds heavenly to me! How can any rational human being believe such rubbish?

18. It is no coincidence that Adam and Eve were punished for eating the fruit from the "Tree of Knowledge" (The suggestion that the "Tree of Knowledge" applied exclusively to knowledge of good and evil is debunked by most contemporary scholars—in the ancient text "good and evil" is a merism, a figure of speech wherein paired opposites mean *all* or *everything*, such as our expression "they searched *high and low."* "The Tree of Knowledge of Good and Evil" meant all knowledge. Thus I prefer to simply drop the confusing "good and evil"; others refer to it more accurately as "The Tree of Knowledge, Both Good and Evil" which may be a more accurate translation of the original Hebrew merism.) Nor is it a coincidence that Satan's other name, Lucifer, literally means *"bringer of light."* By rigorously controlling what you may or may not accept as truth, your religion can control your behavior, not always for the good. The pathetic ignorance of science found among fundamentalist Muslims and Christians reflects their willingness to embrace the irrational and supernatural over the rational and natural in their worldviews. Bertrand Russell observed: *"So far as I can remember, there is not one word in the Gospels in praise of intelligence."* In fact, there are several places where wisdom and intelligence are condemned—consider: *"He respecteth not any that are wise of heart."* (Job 37:24) and *"For I will destroy the wisdom of the wise, and will bring to nothing the understanding of the prudent"* (1 Corinthians 1:19). That sounds like something Donald Trump would say.

19. As recently as February 1990, Cardinal Ratzinger (later Pope Benedict XVI) defended the Church's action regarding Galileo as appropriate. In October 1992, Pope John Paul II expressed regret for how the *"Galileo affair"* was handled, and the following month acknowledged that Galileo was actually correct and exonerated him. As ridiculous as it may seem, that was the first time that the Church properly acknowledged that the earth must rotate around the sun. That's pathetic!

20. The actual scripture reads: *"If a man have a stubborn and rebellious son, which will not obey the voice of his father, or the voice of his mother, and that, when they have chastened him, will not hearken unto them: Then shall his father and his mother lay hold on him, and bring him out unto the elders of his city, and unto the gate of his place; And they shall*

say unto the elders of his city, This our son is stubborn and rebellious, he will not obey our voice; he is a glutton, and a drunkard. And all the men of his city shall stone him with stones, that he die: so shalt thou put evil away from among you; and all Israel shall hear, and fear." Deuteronomy 21:18–21). On a lighter note, some of my more contemporary Jewish friends advise me that their grandparents didn't consider the Jewish fetus to be human until it "graduated from medical school."

21. Ancient Egyptians, Greeks, Romans and Jews commonly used potent botanical agents for birth control and abortion. The extinct plant known to the ancients as Silphium (Genus *Ferula*; possibly *Ferula tingitana*) was such an important part of commerce that it was assigned a specific glyph in the Minoan and Egyptian writing systems and appeared on the face of a Cyrenian silver coin in Hellenic times. The consensus among historians is that Silphium was over-harvested to extinction because of its remarkable effectiveness as an abortifacient and birth control agent. The Romans considered Silphium *"worth its weight in silver."* Of course, the extinction of one herb did not lead to the end of abortion in Biblical times. The amazing combination of cohosh and pennyroyal soon replaced the depleted Silphium, and herbally induced abortions remained important in the Middle East and Mediterranean region. During the first several centuries of the Christian era the main dispensers of herbal medicines were older, single women who were excluded from the developing male-dominated Church. With the emergence of Catholicism, these women were labeled witches and were ultimately persecuted—not for providing medicines, but for acting outside the control of the Church. Later the Church would explicate its bizarre reasoning for the immorality of birth control and justify their practice of burning witches.

I should probably note that one of the supposedly rational arguments advanced today against abortion is that women frequently experience *"mental health problems"* following an abortion. That is not entirely accurate. It is true that women who have unwanted pregnancies tend to be less emotionally stable on average and by extension, women seeking abortions are less stable as well. However, their need for psychiatric support after an abortion is no greater than their need was before they became pregnant. In fact, the risk of postpartum depression following a completed pregnancy is far greater than the risk of psychiatric illness after an abortion.

22. We know that Texas attorneys are suckers for conspiracy theories but Houston theologian/attorney Daniel Shea believes that Ratzinger was elected Pope because the College of Cardinals knew that as Pope Benedict XVI, a "head of state," he would be diplomatically immune to pending American lawsuits examining his role in sheltering pedophile priests. It is known that on May 18, 2001, Cardinal Ratzinger authored, signed and sent an official Vatican letter to every Catholic bishop outlining the accepted procedure for dealing with clerical sex abuse cases. Among other things, the letter ordered everyone involved, under threat of excommunication, to keep the evidence confidential for ten years after the victims reached adulthood. The letter was co-signed by Cardinal Tarcisio Bertone, who later went on the record: *"In my opinion, the demand that a bishop be obligated to contact the police in order to denounce a priest who has admitted the offense of pedophilia is unfounded."* That sure sounds like a conspiracy to obstruct justice.

It is worth noting that Pope Benedict's brother, Georg Ratzinger, directed the renowned Regensburg Domspatzen boys' choir from 1964 to 1994. There are over 500 claims of sexual abuse by former choir members of that era. Ratzinger, now in his nineties, denies knowledge of any sexual abuse but admits that he often punished boys by slapping them.

23. Christian indoctrination almost always involves some absolutely horrid, immoral forms of coercion, most famously, the threat of the "Lake of Fire." Many of us learned as little children that if we failed to fully embrace Jesus and did not follow, *"as lambs,"* all the teachings of the Church, our names would *"not be written into the Book of Life"* and upon our deaths we would be *"cast into a lake of fire"* (Revelation 20:15). Actually, the "Book of Life" manipulation did not originate with Christianity—it was found in earlier cultures where the civil/religious authorities actually maintained the list of community members—a sort of civil "book of life." Disobedient members would have their names expunged from the list and were ostracized with the presumption that they would suffer a terrible fate upon their death. Christians may not have invented "The Book of Life" coercion, but they certainly raised it to an art form.

24. Sometimes it is 72, other times it's only 70—but who's going to quibble over two virgins? (Perhaps the two virgins?) Apparently, women will remain as virtual slaves in Muslim Paradise. Recently a scholar of ancient Semitic languages, who wishes to remain anonymous to avoid religious retribution, suggested that the martyr's reward of virgins is likely the result of a mistranslation. The interpretation as *"virgins"* is from the Arabic *"hur,"* transliterated as *"houris"* which is literally *"white ones."* But the passages describing paradise in the Koran make frequent use of the word *"hur"* to mean *"white raisins."* White raisins were a great delicacy in the ancient world. If this is so, we can only imagine the surprise of the tattered and mutilated young martyr who receives only a handful of raisins when he arrives in paradise. It is an amusing conjecture, but probably not accurate—there are far too many other references in the Koran that speak of the *"dark-eyed, high-bosomed virgins"* who will pleasure Mohammed's followers. In addition, the Koran also advises that paradise will have *"handsome boys of perpetual freshness."* Mohammed apparently understood the adage that *"sex sells"* and used it as a recruiting tool.

25. If you believe that my life belongs to your god, you may feel that you must prevent me from ending it prematurely even if I'm dying of a painful disease. "Good Christians" would subject me to pointless medical treatment to prolong my suffering and/or maintain me in a "brain-dead" comatose state for years. If some one offers to help me end my life, in most states, he or she will be charged with murder for performing a merciful act. Dr Jack Kevorkian served over eight years in prison after being convicted of second-degree murder for assisting patients to end their lives. "Dr. Death's" rusted-out 1968 Volkswagen microbus where he performed "assisted suicides" was up for sale on eBay in April 2010. Regrettably, eBay required the seller to terminate the auction because the sale violated their terms of use. EBay apparently has a rule prohibiting sale of *"murderabilia that is associated with notorious killers within the last hundred years."* Murderabilia? Notorious killers? Get a life, eBay staff!

26. Genesis 1:28 says: *"And God blessed them. And God said to them, 'Be fruitful and multiply and fill the earth and subdue it and have dominion over the fish of the sea and over the birds of the heavens and over every living thing that moves on the earth.'"* To many

environmentalists this Scripture suggests that man has free rein to do whatever he wishes with the planet's resources.

27. In Genesis 38:8–10 we learn that after God killed Er, he commanded Er's brother Onan to serve as surrogate spouse to Er's wife: *"'Go in to your brother's wife and perform the duty of a brother-in-law to her; raise up offspring for your brother.' But since Onan knew that the offspring would not be his, he spilled his semen on the ground whenever he went in to see his brother's wife, so that he would not give offspring to his brother. What he did was displeasing in the sight of the Lord, and He put him to death also."* The term *onanism* is sometimes used to refer to *coitus interruptus*, historically the most common means of birth control among the poor. In addition, Christian churches have often referred to masturbation as the *"the sin of Onan."*

28. While various Judeo-Christian religions may enumerate the Commandments differently, they all stipulate ten moral imperatives that their adherents must follow. Have you ever noticed that the first four of the Biblical Ten Commandments deal exclusively with honoring God? Is he really that insecure? Apparently so—the only "unforgivable sins" in the Christian Bible involve blasphemy (see Matthew 12:31—*"whoever shall speak against the Holy Spirit, it shall not be forgiven him"*— also see Luke 12:10 and Mark 3:29 for more threats). The fifth Commandment is concerned with honoring one's parents, which under most circumstances is moral but is not invariably so! That leaves just five Commandments for things that are actually immoral—homicide (#6), adultery (#7), stealing (#8), lying about others (#9) and covetous materialism (#10). It makes me wonder about the value system and intent of the original author(s).

If you were raised in the historically idolatrous Catholic Church you may never have learned the original Second Commandment that prohibits idolatry (*"You shall not make for yourself an idol in the form of anything in heaven above or on the earth beneath or in the waters below. You shall not bow down to them or worship them; for I, the LORD your God, am a jealous God, punishing the children for the sin of the fathers to the third and fourth generation"* [Exodus 20:4-6].). The Catholic Church deleted it, but in order to keep the number at ten (God apparently counts on his fingers, too), they split the Biblical Tenth (*"Thou shalt not covet thy neighbour's house, thou shalt not covet thy neighbour's wife, nor his manservant, nor his maidservant, nor his ox, nor his ass, nor any thing that is thy neighbour's*") into two different Commandments and clearly changed the implication of coveting *"thy neighbor's wife"* from materialistic to sexual. Actually a similar argument can be made regarding Number Seven on adultery. The original purpose of that prohibition was to prevent women from cheating and bearing illegitimate heirs. The ancient Jews had little problem with "requiring" a recent widow without sons to sleep with the brothers of her deceased spouse in order to bear genetically appropriate heirs to his estate. Women were considered property, and the original intent of Commandment Seven was materialistic, not sexual. Women were often stoned to death but men were rarely punished for adultery. The early Christian Church did little to improve the status of women but did expand the list of "sex crimes."

29. Every time I hear someone refer to their god as the "Supreme Being" I have trouble not laughing—invariably I think of the movie *Time Bandits* and recall little Kevin not-so-meekly addressing the Supreme Being, played by Sir Ralph Richardson. If you like mildly sacrilegious humor you should see this under-appreciated 1981 film written by Terry Gilliam and Michael Palin.

30. Please don't infer that I object to homosexuals in the priesthood or that I equate homosexuality and pedophilia. I'm certain that many gay men become kind and sensitive priests. Homosexuals are rarely pedophiles—pedophiles are mostly individuals who never developed the capacity for intimate relationships with other adults—neither men nor women—and prefer children. Most pedophiles are able to control their impulses, but others, finding themselves in a position of power, may sexually exploit children. Unfortunately, the media often uses the term *"homosexual child abuse"* where the word "homosexual" refers to the relationship between victim and perpetrator (boy and man) and not the sexual orientation of the perpetrator. The hatred of gays has been exacerbated by this careless terminology.

Homosexual men have often constituted a significant percentage of the Catholic clergy. Estimates for gays in the priesthood range from 30–60%. One recent survey gave the figure of 33%. In an effort to deal with what they saw as a problem, the Catholic Church in the late 20th century distinguished *"homosexual orientation"* from *"homosexual genital activity,"* forbidding the latter while tolerating the former. The Church is still struggling to sort out the relationship between celibacy, homosexuality and pedophilia. It's pretty simple to most rational and educated people—pedophilia is an abusive and controlling act perpetrated by immature and selfish individuals against younger and weaker persons. The sexual orientation of the perpetrator is immaterial. The Church has a long way to go—in April 2010 the Vatican Secretary of State, Cardinal Tarcisio Bertone, blamed gay priests for the pedophilia problem! A few days later the Vatican backpedaled and attempted to apologize for his stupid and thoughtless remarks. Bertone's appointment as Vatican Secretary of State by Benedict XVI surprised a lot of people, but insiders claim that Bertone had been influential in getting Cardinal Joseph Ratzinger elected Pope. Bertone didn't have the credentials one would expect of the Vatican Secretary of State but he did know all of Ratzinger's secrets from their close association during many years of pedophile cover-ups.

31. *"Jesus will fix you, you filthy scum."* This wonderfully terse promise of divine retribution was a sample of Christian hate mail sent to the late Madalyn Murray O'Hair, that infamous and obnoxious atheist who is remembered for successfully challenging school prayer in the US Supreme Court. Murray O'Hair enjoyed baiting her Christian adversaries, and would read her hate mail on the air and ridicule the authors. That one particular quote struck me as so emblematic of pathetic Christian hatred that I recorded it, and it still amuses me to this day. *Life* magazine labeled her *"the most hated woman in America."* Following the Supreme Court ruling against school prayer she was forced to leave Baltimore after her home was repeatedly stoned and vandalized, and her son's pet kitten was strangled, presumably by her "good Christian" neighbors. In August 1995 when she was living in Austin, she and her son Jon and granddaughter Robin disappeared under unusual circumstances—their breakfast dishes were still on the table, and O'Hair had left without her diabetes medication and had made no arrangements for the care of her beloved dogs. Despite solid leads developed by the local media, inquiries by her estranged son William, and pressure from federal authorities, the Austin Police Department chose not to investigate the disappearances until forced to do so by the FBI. The murderer, a former employee, was eventually apprehended after six years of "stonewalling" by the Austin Police. Apparently their indifference to the disappearance

of three atheists was acceptable to the Christian majority in that community. Even in death, Madalyn Murray O'Hair found little sympathy. *"Jesus sure fixed you, Madalyn."*

32. If the death-denying strategy of your culture (e.g., Fundamentalist Christianity or Orthodox Judaism) directly contradicts an alternative worldview (e.g., fundamentalist Islam) you should not be surprised when extremist representatives of that other culture seek to destroy important symbols of your culture (e.g., the World Trade Center and the Pentagon). Nearly 3,000 innocent Americans died in the 9/11 attacks. Sadly, the American public still fails to comprehend the reason for the attacks, and we continue to assume we are entirely blameless. In our great naïveté we are stunned when we are attacked for what we believe, but Allah apparently sees things differently. Of course, Americans, like everyone else, get angry when attacked and we seek retribution. Four days after 9/11, Balbir Singh Sodhi, a turban-wearing Sikh, was gunned down outside his gas station in Mesa, AZ. Sodhi was neither Muslim nor Arab. His killer spent time in a bar before the shooting bragging that he was going to *"kill the ragheads responsible for September 11."* Waqar Hasan of Dallas, TX, a Pakistani convenience store owner, was murdered the same day. Three weeks later Hasan's murderer claimed a second victim, Vasudev Patel, an Indian immigrant who owned a Shell service station in Mesquite, TX. Neither man was Muslim nor Arab. A white supremacist named Mark Stroman was arrested and proudly announced: *"I did what every American wanted to do after September 11th but didn't have the nerve."* On July 20, Stroman was executed by the Great Christian State of Texas despite the tireless efforts of one of his other victims. Rais Bhuiyan, a Muslim immigrant from Bangladesh, survived a load of buckshot to the face but spent several years appealing to the Texas judicial system to spare Stroman's life. Bhuiyan argued, *"If I can forgive my offender who tried to take my life we can all work together to forgive each other and move forward and take a new narrative on the 10th anniversary of 11 September."* The irony that a Muslim asked the State of Texas to "turn the other cheek" was lost on the local authorities and media. Recent Texas governors, seemingly selected on the basis of their mock religiosity and real stupidity, exhibit a special fondness for executions. They fail to acknowledge the similarity between their "Christian" morality and the Sharia law that they dread. Ironically, Mark Stroman was resigned to his fate and was grateful to Rais

Bhuiyan for his efforts—he told reporters, *"It is due to Rais' message of forgiveness that I am more content now than I have ever been."* In his final moments he said, *"Hate is going on in this world and it has to stop—hate causes a lifetime of pain. God bless America. God bless everyone."* And lastly, emulating Gary Gilmore, *"Let's do this damn thing."*

At least five other murders of *"Middle Eastern-appearing"* men were linked to *"hate backlash"* in the month following the 9/11 attacks. In the year that followed at least 80 crimes against individuals and property were shown to be tied to 9/11 backlash. There were likely many more.

On October 7, 2001 the US and British Armed Forces launched Operation Enduring Freedom in response to the 9/11 attacks with the goal of dismantling Al-Qaeda and removing the Taliban from power in Afghanistan. *"My god is better than your god and we'll kick your ass to prove it."* I don't mean to suggest that I disapprove of our actions—I just think it is important that we fully appreciate our own motivation and understand that the hatred engendered by competing world views is deeply rooted in our culture. Despite my religious neutrality, like most other Americans, I was outraged when my country was attacked. Summoned by colleagues to the Biology Department Conference Room TV that morning, I saw the second hijacked plane strike the second Tower and then endured the mind-numbing, repeated reruns of that and the rapidly surfacing videos of the first attack. Our President and our media pundits had no idea what was happening!

Another example of worldview conflict was seen in the summer of 2010 when the Rev. Terry Jones, pastor of the Dove World Outreach Center in Gainesville, FL, proclaimed that 9/11/2010 would be recognized as *"International Burn a Koran Day."* Jones received requests from hundreds of rational people, including President Obama, to not burn the Islamic holy book and cease his hatemongering. He also received hundreds of death threats that, according to his son Luke, did more to dissuade him than any reasonable arguments. But Jones had started something he couldn't stop. He continued to receive numerous letters of encouragement, financial support, and donated Korans from what I like to call the "hateful-faithful." On March 20, 2011, his inflated ego compelled him to actually go through with his threat, and he and 30 members of his congregation held a mock trial, condemned the Koran, soaked a copy in kerosene and proceeded to burn it. I can only imagine that they marched around the flames while singing "Onward Christian Soldiers." Fortunately, no photos or videos of the event were released to the public. Still, the word spread, and by early April 2011 hundreds of thousands of Muslims around the world protested violently. Dozens of innocent people died, including seven UN peacekeepers in Afghanistan. Jones subsequently staged a protest outside the Islamic Center of America in Dearborn, MI on Good Friday, April 22, 2011. He was arrested, posting $1 bond with the provision that he leave town and not return within three years.

33. I am greatly enamored with Becker's reasoning and central arguments, but I find much of his book a bit dull and dated. Despite being trained as a cultural anthropologist at Syracuse, Becker felt compelled to dwell on psychoanalytic theory in his book. His modest discussion of comparative cultures and his limited grasp of evolutionary biology and neuroscience reduce the effectiveness of his arguments for the contemporary reader. Had Becker lived a few more years he would have benefited from the intellectual renaissance in psychology that sprang from human sociobiology and modern evolutionary theory. But it was a very impressive piece of work for 1973.

34. I should also concede that belief probably eases the proximate transition from life to death—the final moments in the actual event of dying. I've known of individuals who, in their final moments, spoke of being greeted by long-deceased loved ones. I won't get into a discussion of the physiological events in the oxygen-starved dying brain and how hallucinations are a common effect. But hallucinations or not, if they bring a smile to the face of the dying person and comfort the mourners they are not all that bad. We would expect that a shared familial death denial scheme would also alleviate the grief of surviving family members following death of a loved one —in most cases they should be confident of meeting the deceased again in the afterlife. But it has not been my experience that the surviving family members enjoy the benefit of reduced grieving. Conversely, we would expect nonbelievers to experience the greatest grief upon loss of a loved one, and that, as well, has not been my experience. This apparent paradox warrants further analysis. As noted in the text, a recent (March 2017) comprehensive study in *Religion, Brain and Behavior* suggests that both very religious individuals and nonbelievers

exhibit less fear of death than less religious individuals. That is much like what we saw with happiness—people secure in either belief or disbelief are generally more happy that those poor ambivalent souls!

35. While I would like to blame all war on religion, that would be simplistic. But I would point out that war was essentially unknown among our animistic hunter-gatherer ancestors and it did not become widespread until 4–6 thousand years ago coinciding with the emergence of monotheism. Following the Neolithic Revolution, the cohesion of larger population groups was facilitated by in-group cooperation and its reciprocal out-group hostility. Religions have normally been the major contributors to out-group aggression—i.e., war.

36. If you don't believe that the Great World Wars of the 20th century were largely worldview conflicts, I urge you to study history a little more carefully. "Aryan superiority," "Russian pan-Slavic nationalism," "Japanese imperialism," etc. are all examples of unifying cultural-religious worldviews that promoted aggression toward "out-groups."

37. Irons argues that *"the most powerful cultural signals of commitment are religious ones, and thus evolution has built into human beings a strong propensity to seek religious orientation toward life and to hold this orientation to be of the highest value."* That last phrase may be an exaggeration, but Irons is fundamentally correct. If he is willing to acknowledge "reason and science" as a form of "religious orientation," I won't argue with him. This is an area that warrants further study. See Chapt. 13, "Religion as a Hard-to-Fake Sign of Commitment," in *Evolution and the Capacity for Commitment*, (Randolph Nesse, ed., Russell Sage Foundation, NY, 2001).

38. Empathy is, of course, the capacity to recognize feelings of others. We often mistakenly assume that it is a uniquely human behavior, but anyone who has ever owned a dog knows better. Dogs and many primates are able to empathize across species and recognize feelings in their human companions. Rudimentary empathic behavior has been observed in a variety of mammals including non-primates. In human children the capacity to recognize the feelings of others usually begins by age two. At about the same age they begin to assess other's intentions. By three or four years most children will exhibit sympathy and compassion—that is, they are

able to share the subject's suffering. These are genetically programmed behaviors that are shaped by experience. Many psychologists agree that prosocial behavior, particularly altruism, is frequently motivated by sympathy. Simple altruism occurs when one individual sacrifices on behalf of another. Altruism is often kin-directed—a mother's sacrifice on behalf of her children is altruistic. The individual performing the act of seeming self-sacrifice may actually gain fitness—the sacrifices by the mother on behalf of her offspring can increase the proliferation of her own genes through her children. Thus, the evolution of such behavior is easily explained by natural selection. Direct reciprocity involves sacrificing on behalf of others with anticipation of repayment at a later date—it is widespread in the animal kingdom and can enable the evolution of social cooperation. Our earliest hunter-gatherer ancestors lived in small groups and were able to accurately monitor exchanges for fairness. Non-reciprocators were easily identified and others in the group could refuse to exchange with a "cheater." Individual "reputation" was important to success (and biological fitness). Mate choice, an aspect of sexual selection, was influenced by familial and individual reputations for altruism, cooperation and fairness. As social group size increased during the Neolithic revolution, it became desirable to transact with strangers beyond one's immediate kinship group, and indirect reciprocity became essential. If direct reciprocity is popularly explained as "I'll scratch your back if you scratch mine," indirect reciprocity is more like: "I'll scratch your back if there is a high probability that someone else will come along later and scratch mine." Indirect reciprocity may be a little more complex but is still attributable to natural selection. Its exact roots are unclear—it would seem maladaptive to treat strangers with the same generosity that one treats family. Some researchers have suggested that the kin-directed altruism found throughout the primate series became misdirected as societies grew in size. The Stone-Age human brain had not been required to make distinctions between kin and non-kin because it rarely encountered strangers—thus, the latter were treated the same as kin. The primitive human brain couldn't comprehend meeting someone it would never encounter again, so it treated strangers as part of its family. Whatever the roots, fairness between strangers allowed certain groups to cooperate successfully and thrive, whereas groups comprising more selfish individuals failed. Social norms and institutions (e.g., markets and religions) that promoted fairness are viewed by many as inevitable evolutionary steps. "Reputation" became a crucial element, and those institutions that could verify and vouch for reputation became central to the society. Norms of fairness that encouraged transactional honesty and punished "cheaters" were favored by both biological and cultural evolution. Over thirty years ago Richard Alexander (*The Biology of Moral Systems*, Aldine De Gruyter, NY, 1987) cogently argued that much of human moral systems derive from indirect reciprocity. It may seem peculiar, but recent research data support the view that fairness norms are to a large extent the product of biological evolution. While there is some cultural and individual variation, there is a universal human tendency to act more fairly (generously) than economic models predict. I have written about the unique nature of "human fairness" in somewhat greater detail in Chapter XII, "My Naturalistic World View," in *A Fox Family History - Book Two* (2015).

39. In an elegant cross-cultural study entitled "The Negative Association between Religiousness and Children's Altruism across the World" (*Current Biology*, 5 Nov 2015) Jean Decety and co-workers demonstrated that children reared in various religious households were significantly *less* empathic than children reared in atheistic households. On an arbitrary scale of "generosity," Muslim children's scores averaged 3, Christian children averaged 3.5, and atheist children averaged 4—the children of atheistic households were significantly more generous than children of both Christian and Muslim households. Similarly, children in religious households tended to be more judgmental and punitive in dealing with perceived moral transgressors. These differences between religious and atheist children increased with age. Not surprisingly, the parents of children reared in religious homes incorrectly assessed their own children as morally superior: *"Across all countries, parents in religious households reported that their children expressed more empathy and sensitivity for justice in everyday life than non-religious parents. However, religiousness was inversely predictive of children's altruism and positively correlated with their punitive tendencies. Together these results reveal the similarity across countries in how religion negatively influences children's altruism, challenging the view that religiosity facilitates prosocial behavior."*

The authors summarize their findings thus: *"Overall, our findings cast light on the cultural input of religion on prosocial behavior and contradict the common-sense and popular assumption that children from religious households are more altruistic and kind toward others. More generally, they call into question whether religion is vital for moral development, supporting the idea that the secularization of moral discourse will not reduce human kindness—in fact, it will do just the opposite."*

Read that again—*"they call into question whether religion is vital for moral development"..."secularization of moral discourse will not reduce human kindness—in fact, it will do just the opposite."*

SUBJECT INDEX

PERSON INDEX

About the Author

Kevin A. Fox, PhD is a retired Distinguished Teaching Professor of Biology (SUNY Fredonia). His professional scholarship and publications have embraced a wide range of interests from sociobiology (evolutionary biology and animal behavior) to physiology (neurobiology and psychopharmacology). This book reflects his current interest in "neurotheology"—the study of the neurological bases of theological belief. As a "devout non-believer" he has sought to comprehend the evolutionary and cultural roots behind the emergence and persistence of humankind's theistic religions. Global circumstances—increasing population, diminishing resources and declining environmental quality—presage escalating global strife. Unfortunately, there is little evidence that extant world religions will do anything other than exacerbate that strife. It is urgent that we either understand and refocus our prevailing religions, or we jettison them in favor of rational and authentically altruistic (and probably non-theistic) moral systems.

Other recent books by the author that probably won't interest you include:

A Fox Family History: Family History and Genealogy for the Descendants of Archer and Mabel Fox of Wilton, NH Book One
Fox Editing & Publishing, 2015—Second edition

A Fox Family History: Family History and Genealogy for the Descendants of Archer and Mabel Fox of Wilton, NH Book Two—The Wilton Years
Fox Editing & Publishing, 2015

Both available online at: https://foxediting.com/kevin-a-fox/

FOX

San Francisco
FOXEDITING.COM